Lecture Notes in Computer Scie

T0237969

Commenced Publication in 1973
Founding and Former Series Editors:
Gerhard Goos, Juris Hartmanis, and Jan van Leeuwen

Editorial Board

David Hutchison
Lancaster University, UK

Takeo Kanade
Carnegie Mellon University, Pittsburgh, PA, USA

Josef Kittler
University of Surrey, Guildford, UK

Jon M. Kleinberg
Cornell University, Ithaca, NY, USA

Alfred Kobsa
University of California, Irvine, CA, USA

Friedemann Mattern
ETH Zurich, Switzerland

John C. Mitchell
Stanford University, CA, USA

Moni Naor
Weizmann Institute of Science, Rehovot, Israel

Oscar Nierstrasz
University of Bern, Switzerland

C. Pandu Rangan
Indian Institute of Technology, Madras, India

Bernhard Steffen
University of Dortmund, Germany

Madhu Sudan
Microsoft Research, Cambridge, MA, USA

Demetri Terzopoulos
University of California, Los Angeles, CA, USA

Doug Tygar
University of California, Berkeley, CA, USA

Gerhard Weikum
Max-Planck Institute of Computer Science, Saarbruecken, Germany

Richard Fuller   Xenofon D. Koutsoukos (Eds.)

Mobile Entity Localization and Tracking in GPS-less Environments

Second International Workshop, MELT 2009
Orlando, FL, USA, September 30, 2009
Proceedings

 Springer

Volume Editors

Richard Fuller
Gilroy, CA, USA
E-mail: fuller.richard@gmail.com

Xenofon D. Koutsoukos
Vanderbilt University
EECS Department/ISIS
362 Jacobs Hall 400, 24th Avenue South
Nashville, TN 37235, USA
E-mail: Xenofon.Koutsoukos@vanderbilt.edu

Library of Congress Control Number: 2009934296

CR Subject Classification (1998): C.3, I.5.4, F.2.2, I.3.5, C.2.1, C.2

LNCS Sublibrary: SL 3 – Information Systems and Application, incl. Internet/Web and HCI

ISSN   0302-9743
ISBN-10   3-642-04378-X Springer Berlin Heidelberg New York
ISBN-13   978-3-642-04378-9 Springer Berlin Heidelberg New York

This work is subject to copyright. All rights are reserved, whether the whole or part of the material is concerned, specifically the rights of translation, reprinting, re-use of illustrations, recitation, broadcasting, reproduction on microfilms or in any other way, and storage in data banks. Duplication of this publication or parts thereof is permitted only under the provisions of the German Copyright Law of September 9, 1965, in its current version, and permission for use must always be obtained from Springer. Violations are liable to prosecution under the German Copyright Law.

springer.com

© Springer-Verlag Berlin Heidelberg 2009
Printed in Germany

Typesetting: Camera-ready by author, data conversion by Scientific Publishing Services, Chennai, India
Printed on acid-free paper   SPIN: 12760473   06/3180   5 4 3 2 1 0

Preface

This volume contains the proceedings of the Second International Workshop on Mobile Entity Localization and Tracking in GPS-less Environments (MELT 2009), held in Orlando, Florida on September 30, 2009 in conjunction with the 11th International Conference on Ubiquitous Computing (Ubicomp 2009). MELT provides a forum for the presentation of state-of-the-art technologies in mobile localization and tracking and novel applications of location-based services. MELT 2009 continued the success of the first workshop in the series (MELT 2008), which was held is San Francisco, California on September 19, 2008 in conjunction with Mobicom.

Location-awareness is a key component for achieving context-awareness. Recent years have witnessed an increasing trend towards location-based services and applications. In most cases, however, location information is limited by the accessibility to GPS, which is unavailable for indoor or underground facilities and unreliable in urban environments. Much research has been done, in both the sensor network community and the ubiquitous computing community, to provide techniques for localization and tracking in GPS-less environments. Novel applications based on ad-hoc localization and real-time tracking of mobile entities are growing as a result of these technologies. MELT brings together leaders from both the academic and industrial research communities to discuss challenging and open problems, to evaluate pros and cons of various approaches, to bridge the gap between theory and applications, and to envision new research opportunities.

The research contributions in these proceedings cover significant aspects of localization and tracking of mobile devices that include techniques suitable for smart phones and mobile sensor networks in both outdoor and indoor environments using diverse sensors and radio signals. Novel theoretical methods, algorithmic design and analysis, application development, and experimental studies are presented in 14 papers that were reviewed carefully by the program committee. In addition, three invited papers, with topics on location determination using RF systems, Cramér-Rao-Bound analysis for indoor localization, and approaches targeting mobile sensor networks, are also included in the proceedings.

We would like to thank the authors of the submitted papers, the Program Committee members, the additional reviewers, and the workshop organizers for their help in composing a strong technical program. We also thank the Ubicomp 2009 organizing committee for providing a premier outlet for the workshop. Finally, we would like to thank Springer for having agreed to publish these proceedings as a volume in the Lecture Notes in Computer Science series.

July 2009

Richard Fuller
Xenofon Koutsoukos
Ying Zhang

Organization

Organization Committee

Program Co-chairs

Richard Fuller
(Wireless Communications Alliance, USA)
Xenofon Koutsoukos
(Vanderbilt University, USA)

Publication Chair   Ying Zhang (Palo Alto Research Center, USA)
Publicity Chair   Andras Nadas (Vanderbilt University, USA)
Web Chair   Isaac Amundson (Vanderbilt University, USA)
Industry Sponsorship Chair   Richard Fuller
(Wireless Communication Alliance, USA)

Technical Program Committee

Romit Roy Choudhury	Duke University, USA
Maurice Chu	Palo Alto Research Center, USA
Richard Fuller	Wireless Communications Alliance, USA
Ismail Guvenc	DoCoMo USA Communications Lab, USA
Martin Griss	CMU West, USA
Tian He	Univeristy of Minnesota, USA
Pan Hui	Deutsche Telekom Laboratories, Germany
Lukas Kencl	Czech Technical Univ., Prague, Czech Republic
Xenofon Koutsoukos	Vanderbilt University, USA
Branislav Kusy	Stanford University, USA
Juan Liu	Palo Alto Research Center, USA
Yannis Paschalidis	Boston University, USA
Kurt Partiridge	Palo Alto Research Center, USA
Bodhi Priyantha	Microsoft Research, USA
Zafer Sahinoglu	Mitsubishi Electric Research Lab, USA
Cem Saraydar	General Motors, USA
Yi Shang	University of Missouri, USA
Radu Stoleru	Texas A&M University, USA
Vincent Tam	Hongkong University, China
Hui Zang	Sprint, USA
Ying Zhang	Palo Alto Research Center, USA

Additional Referees

Isaac Amundson   Manish Kushwaha
Katerina Dufkvoá   Cigdem Sengul
Michal Ficek   Ziguo Zhong

Table of Contents

Localization by Experiments

Invited Papers

MGALE: A Modified Geometry-Assisted Location Estimation Algorithm Reducing Location Estimation Error in 2D Case under NLOS Environments

Pampa Sadhukhan[1] and Pradip K. Das[2]

[1] School of Mobile Computing & Communication,
Jadavpur University, India 700032
pampa.sadhukhan@gmail.com
[2] Faculty of Engineering & Technology,
Mody Institute of Technology & Science, India 332311
pkdas@ieee.org

Abstract. Positioning of a Mobile Station (MS) is mandatory for implementing emergency call services and providing location-based services. The major problems facing accurate location estimation of the MS are Non-Line-of-Sight (NLOS) propagation and hearability. Thus several location algorithms focusing on NLOS mitigation and addressing hearability problem have been studied. Among these, Geometry-Assisted Location Estimation (GALE) algorithm based on three TOA measurements can estimate the 2D location of the MS with reasonable precision. However 2D GALE scheme fails to meet Federal Communication commission (FCC) requirement for phase 2. Thus, in this paper, we have proposed a modification over 2D GALE algorithm with three TOA measurements to meet the FCC target while preserving the computational efficiency of GALE scheme. Our proposed algorithm considers the geometry of the TOA measurement circles rather than the standard deviation of the TOA measurements. Simulation results show that our algorithm provides better location accuracy compared to 2D GALE algorithm.

Keywords: Location Estimation, Base Station (BS), Mobile Station (MS), Geometry-Assisted Location Estimation Algorithm (GALE), Time-of-Arrival (TOA), Non-Line-Of-Sight (NLOS).

1 Introduction

Wireless location estimation has drawn a considerable attention from the .. researchers over the past few decades. Positioning Technologies that help to estimate the position of a Mobile Station (MS) are not only mandatory for Emergency 911 (E-911) call services [1], but also applicable for providing location-based services, route guidance, vehicle tracking. Global Positioning System (GPS) is the most accurate and viable solution to providing the aforementioned services. However, the high expenses of GPS receiver and technical challenges associated

R. Fuller and X.D. Koutsoukos (Eds.): MELT 2009, LNCS 5801, pp. 1–18, 2009.
© Springer-Verlag Berlin Heidelberg 2009

with replacing all existing cellular handsets with GPS-equipped handsets have motivated researchers to work with radio location systems.

Several location estimation techniques widely employed in radio location systems are based on Received Signal Strength (RSS), Time-Of-Arrival (TOA), Time-Difference-Of-Arrival (TDOA) and Angle-Of-Arrival (AOA). The rapidly changing signal propagation condition between the BS and the MS makes RSS-based location technique obsolete in the outdoor environment. Among the network-based localization techniques, TDOA and TOA based location technique require at least three properly located Base Stations (BSs), whereas AOA based technique requires only two BSs at minimum to determine 2D location of a MS [2].

A major issue in providing accurate location estimation of the MS is that the signal measurement generally includes error due to Non-Line-Of-Sight (NLOS) propagation in the urban and metropolitan area and also the error due to system measurement noises. However, existence of NLOS errors dominates the errors due to system measurement noises. Thus traditional algorithms based on TDOA, TOA and hybrid TDOA/AOA technique proposed in [3]-[6], would fail to obtain the MS's location estimate with desired accuracy as these algorithms have been designed to work under Line-of-Sight (LOS) environment including small measurement errors. A considerable amount of research has been done in the field of identifying and mitigating the NLOS error based on the assumption that presence of NLOS error makes the range measurement greater than LOS measurement.

The approach for alleviating NLOS error proposed in [7] requires a time-series of range measurements from each BS involved over a time span of few seconds. It can identify the presence of NLOS error in a range measurement only if the set of range measurements used for location purposes include at least one LOS range measurement. Moreover, it depends on the prior knowledge of standard deviation of measurement noise to reconstruct the LOS measurement value from NLOS corrupted range measurement. There are several approaches [8]-[10] that can mitigate the effect of NLOS errors by considering the range residual between the measured range and estimated range while using only a single measurement at each BS. These approaches fail to provide desired location accuracy in case only the NLOS propagation exists between the MS and the BSs.

Another issue in network-based location estimation scheme is the computational complexity incurred by solving the nonlinear equations associated with the MS's location estimate. The two-step Least Square (LS) method that can provide the Maximum-Likelihood estimate of the MS's position by using only two computing iterations has been adopted in several location estimation algorithms [3], [6], [11]. In [11], authors have shown that TOA-based algorithm can provide better accuracy in the location estimation than that of TDOA-based algorithm [3] even under the presence of NLOS errors. So we limit our discussion to TOA-based location techniques. However, the algorithm considered in [11] performs well in case of more than five receivers are available for location purposes.

Some statistical algorithms based on channel-scattering model such as Ring/ Disk of Scatterers and Gaussian Distributed Scatterer in [12], [17], require only three BSs to estimate the LOS measurements from NLOS corrupted measurements by comparing measured statistics of the TOAs of several multipath signals to the statistics generated from well-known scattering models. However, these methods can be applied in an area if the prior knowledge of the scattering model for that area is available.

Hybrid technique based on TOA/AOA is useful in situations where hearability problem exists and it can perform well when at least the serving BS is at LOS with the MS. The location scheme based on hybrid TOA/AOA technique [14] has adopted a non-linear constrained optimization procedure with bounds on errors incorporated in range and angle measurement inferred from geometry (HTA) and also a least-square solution of lines of position derived from both range measurement and angle measurement equations (HLOP). The Range Scaling Algorithm (RSA) [13] based on three TOA measurements also adopts a constrained nonlinear optimization procedure for estimating the true ranges between the BSs and the MS by defining the true ranges as the scaled version of the measured ranges and utilizes the LLOP algorithm outlined in [5] to obtain the MS's position from the estimated LOS ranges. This approach utilizes the bound on the NLOS error and the geometry of cell layout to compute the value of the scale factors. However, this scheme incurs heavy computational complexity by solving the constrained nonlinear optimization problem and it is actually viable for smaller cells.

In [15], authors have proposed a Kalman-based Interacting Multiple Model (IMM) smoother working with three TOA measurements for efficiently mitigating the NLOS errors under uncertain environmental conditions where the transmission channel between the MS and the BS transits between LOS and NLOS condition. However, location accuracy provided by this approach depends on the values assigned to the two-state transition probability matrix to define the transition of transmission channel between LOS and NLOS mode.

Geometry-Assisted Location Estimation (GALE) algorithm proposed in [16], can estimate the 2D location of the MS with tolerable precision in NLOS environments by adopting the geometric constraints between the MS and the BS within the formulation of two-step LS method. The authors in [16] have shown that the GALE scheme can achieve better precision in 2D location estimation compared to the two-step LS algorithm [11], LLOP scheme [5], and RSA scheme [13] while retaining computational efficiency of two-step LS scheme. It attempts to confine the MS's position within the overlap region between the range measurement circles by considering the standard deviations of the TOA measurements, which generally do not always measure the amount of NLOS errors incorporated into those TOA measurements. GALE scheme fails to acquire reasonable location accuracy in case some TOA measurement taken from a BS with low jitter includes large NLOS error. Moreover, location estimation errors provided by GALE scheme are far above the Federal Communication Commission(FCC) target (for 67% location error at 100m and 95% location error at 300m).

In this paper, we have proposed a modification over 2D GALE algorithm with three TOA measurements to obtain a substantially improved accuracy in 2D location estimation of the MS. Our proposed scheme considers the intersecting point of the common chords among the range measurement circles and the geometric constraints between the position of the MS and the BSs to confine the MS's position within the overlap region between those range circles rather than taking into account the standard deviation of the range measurements. Numerical results demonstrate that our scheme can achieve the required location accuracy defined by FCC for phase 2 and acquires better precision in location estimation compared to GALE algorithm in NLOS environment. The remaining part of this paper is organized as follows. Section 2 describes the GALE algorithm for 2D location estimation in brief. In section 3, we present our algorithm Modified Geometry-Assisted Location Estimation Algorithm (MGALE). Section 4 evaluates the performance of the proposed scheme to show its effectiveness and to compare it with 2D GALE scheme. Section 5 draws the conclusion and presents our future goal.

2   2D GALE Algorithm with Three TOA Measurements

2D GALE scheme is based on the proposition that the position of MS should always fall inside the overlap region between the circles drawn from TOA measurements r_1, r_2 and r_3 as shown in fig.1. To confine the MS's expected

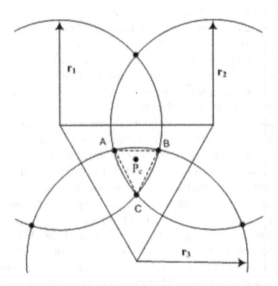

Fig. 1. The schematic diagram of the 2D TOA-based location estimation for NLOS environments

position within the area of $\triangle ABC$ (fig.1), GALE scheme introduces the following constraints.

The virtual distance γ between the MS's position and the three intersecting points A, B, C (fig.1) is defined as

$$\gamma = \left[\sum_{k=1}^{m} \frac{1}{m} \|P - P_k\| \right]^{\frac{1}{2}}, \tag{1}$$

where $P = (x, y)$ represents the MS's position and P_k represents the intersecting points around the overlap region that is points A, B, C and m denotes the number of intersecting points around overlap region.

Similarly the expected virtual distance is defined as

$$\gamma_e = \left[\sum_{k=1}^{m} \frac{1}{m} \|P_e - P_k\| \right]^{\frac{1}{2}} = \gamma + n_\gamma, \tag{2}$$

where P_e denotes the expected position of the MS that is determined by considering the standard deviations from three TOA measurements. The major objective of the GALE scheme is to minimize the deviation between the virtual distance γ and the expected virtual distance γ_e that is the absolute value of n_γ. The detailed description of the GALE algorithm for 2D location estimation can be found in [16].

3   Proposed MGALE Algorithm for 2D Location Estimation

The proposed MGALE scheme based on only three TOA measurements for 2D location estimation of the MS is presented in this section. Our proposed algorithm takes into account the fact that the presence of NLOS error makes the measured ranges larger than the true ranges. The position of the MS lies within the overlap region among the range circles as shown in fig.1 under the condition that NLOS error is always positive and larger than the measurement noise. It can be observed from fig.2 that the intersecting point of the common chords of the range circles meet at some point within the overlap region among the range circles and it divides that overlap region into three sub-regions. The MS should be located within only one of these sub-regions. The main objective of the proposed MGALE scheme is to find out the sub-region confined by the intersecting point of the common chords of the range circles and another two points among the intersecting points around the overlap region within those circles where the MS should be located. Then it estimates the MS's position taking into consideration the geometry of the range circles rather than the standard deviations of the TOA measurements.

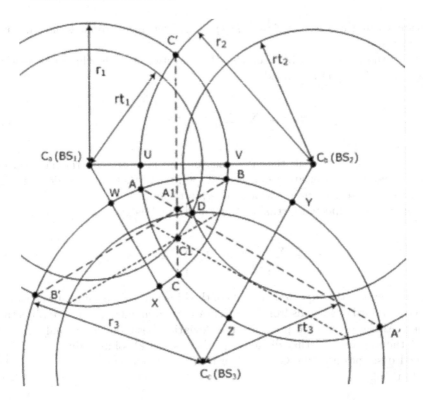

Fig. 2. The schematic diagram of $MS's$ actual position and movement of the intersecting point of three common chords with respect to variation in TOA measurement under NLOS environments

3.1 Determining the Subset of Overlap Region among the Range Circles Where the MS Should Be Located

The proposed MGALE Scheme utilizes the following proposition and several observations from fig. 2 to find out the sub region confined by the intersecting point of the common chords of the range circles and another two points among the intersecting points around the overlap region where the position of the MS should lie.

Proposition 1. *The common chord between the two range circles moves toward the centre of that range circle associated with TOA measurement incorporating least NLOS error away from the MS's position.*

Proof. Fig. 3 shows the movement of the common chord between the two range circles with respect to the position of the MS while varying the NLOS errors incorporated into TOA measurements. Suppose l_1 and l_2 are the true range measurements from the BSs located at points C_a and C_b respectively and the MS is located at point A as shown in fig.3. Now, considering the common chord

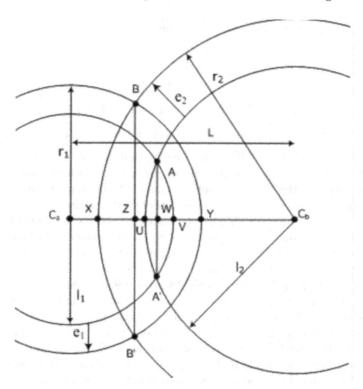

Fig. 3. Geometry of two range circles drawn from TOA measurements and $MS's$ actual position with respect to the common chord between the range circles

between the two circles drawn from range measurements l_1 and l_2 intersects the line adjoining the points C_a and C_b at point W as shown in fig.3, we get,

$$\overline{C_aW} + \overline{C_bW} = L \left(\because \overline{C_aC_b} = L \right).$$
$$\overline{AW}^2 = l_1^2 - \overline{C_aW}^2 = l_2^2 - \overline{C_bW}^2 \left(\because \overline{AA'} \perp \overline{C_aC_b} \right).$$

Rearranging the above formulae and normalizing by $\overline{C_aW} + \overline{C_bW}$, we get

$$\overline{C_aW} = \frac{L^2 - (l_2^2 - l_1^2)}{2L}, \quad \overline{C_bW} = \frac{L^2 + (l_2^2 - l_1^2)}{2L} \tag{3}$$

Let us consider, due to the existence of NLOS error, measured ranges are r_1 and r_2 and the NLOS error incorporated into those range measurements are e_1 and e_2 respectively as shown in fig. 3. Considering the common chord between the two circles drawn from range measurements r_1 and r_2, would intersect the line C_aC_b at point Z and following above equation (3), we get

$$\overline{C_aZ} = \frac{L^2 - (r_2^2 - r_1^2)}{2L}, \quad \overline{C_bZ} = \frac{L^2 + (r_2^2 - r_1^2)}{2L} \tag{4}$$

Now considering $e_1 < e_2$, we get the following constraint

$$r_1 - l_1 < r_2 - l_2$$

$$\Rightarrow r_2 - r_1 > l_2 - l_1$$

$$\rightarrow \left(r_2^2 - r_1^2\right) > \left(l_2^2 - l_1^2\right) \left[\because r_2 + r_1 > l_2 + l_1\right]$$

If the above constraint is satisfied the following relations hold.
$\overline{C_a Z} < \overline{C_a W}$ and $\overline{C_b Z} > \overline{C_b W}$, which means that the common chord moves toward the centre of the range circle drawn from range measurement r_1 away from the position of the MS, if r_1 contains smaller NLOS error than r_2.

Based on the above proposition and several observations from fig. 2, it can be shown that the location of the MS can be found out within the region bounded by the intersecting point of the common chords among the circles and another two vertices of $\triangle ABC$ that lie on the boundary of the circle drawn from range measurement incorporating least amount of NLOS error.

It is observed from fig.2 that if three TOA measurements were free from NLOS errors and measurement noises, the three circles drawn using those three TOA measurements would meet at a single point that is the position of the MS (point D in fig.2). In that case the common chords among those circles would also meet at that point. Here we have assumed (as shown in fig.2) rt_1, rt_2 and rt_3 are the true distances between the MS and BS_1 (Home BS), BS_2 and BS_3 respectively. On the other hand, r_1, r_2 and r_3 are considered, as the measured distances between the MS and BS_1, BS_2 and BS_3 respectively and the intersecting point of the common chords between the range circles would meet at point A1 in that case as shown in fig.2.

For TOA measurement r_1, the MS should be located on the boundary of circle with radius r_1 in case r_1 is free from any kind of noises. As r_1 incorporates some amount of noise, the MS's position moves toward the centre of the circle with radius r_1. Therefore, it can be inferred that the MS should be located nearest to the boundary of the circle drawn from TOA measurement including smallest amount of NLOS error.

The above inference together with the proposition that the common chord between the two range circles moves toward the centre of range circle having smaller NLOS error away from the MS's position, implies the following fact.

The intersecting point of the three chords AA', BB' and CC' moves toward the centre of the circle drawn from the TOA measurement incorporating smallest NLOS error whereas the MS falls very nearer to the boundary of that circle. Thus, the MS's position lies between the intersecting point of the common chords and the boundary of that range circle corresponding to the least NLOS error. The range measurement r_1 includes least amount of NLOS error among the three range measurements as shown in fig.2 and thus the position of MS lies between the intersecting point of the common chords and the boundary of circle associated with r_1 as depicted in fig.2. Hence the MS's position falls within the region confined by points A1, B and C instead of the region confined by points A, B and C.

Similarly, if TOA measurement r_3 is reduced to rt_3, the common chords between the circles would meet at point C1. In that case, the MS lies between the point C1 and the boundary of the circle drawn from the range measurement rt_3 that incorporates least amount of NLOS error as shown in fig. 2.

The above observations lead to the conclusion that *the MS should always fall within the sub region confined by the intersecting point between the common chords among the range circles and another two vertices of $\triangle ABC$ that lie on the boundary of the range circle associated with TOA measurement having smallest amount of NLOS error.*

3.2   Relationship between NLOS Error and BS-MS Range

The estimated TOA value τ can be represented as follows.

$$\tau = \tau_0 + \tau_e, \tag{5}$$

where τ_0 is the true TOA value and τ_e includes NLOS error and system measurement error. In [19] authors have proposed the following method for calibrating NLOS error.

The TOA measurement τ estimated from the first moment of multipath power delay profile can be represented by the following equation.

$$\tau = \tau_0 + \frac{\sum_i \tau_i \cdot P_i}{\sum_i P_i} = \tau_0 + \tau_m, \tag{6}$$

where τ_i is the delay relative to a direct wave, P_i is the corresponding power and τ_m is defined as mean excess delay. Thus, the positive NLOS bias is approximated by the mean excess delay τ_m.

In [18], authors have presented the evidence to the conjecture that the median of the Root-Mean-Square delay spread (τ_{rms}) grows mildly with the distance between the MS and the BS. Moreover, authors in [19], have presented a relationship between τ_m and τ_{rms} of the form $\tau_m \approx k\tau_{rms}$, where k is proportionality constant based on measurement results from Motorola and Ericsson. Thus, results from those papers imply that larger range measurement may contain higher NLOS error compared to some smaller range measurement.

However the NLOS error included in a range measurement, also depends on the environment, that means, more obstacles between the BS and the MS results in higher NLOS error in the TOA measurement. Thus, it is required to identify the TOA measurement that incorporates least amount of NLOS error as it may happen that the smallest TOA measurement includes larger NLOS error than any other TOA measurement. Our proposed scheme provides a method to select the TOA measurement incorporating the least amount of NLOS error in the following subsection.

3.3 Identification of the TOA Measurement Having Least Amount of NLOS Error

Our proposed scheme provides the following method to select the TOA measurement incorporating the least amount of NLOS error.

The parameter XC_i is introduced to denote the distance between the position of ith BS and the intersecting point between the common chords among the range measurement circles. If three TOA measurements are free from any kind of noises, the circles drawn using those three TOA measurements would meet at a single point (the point D as shown in fig. 2) and the common chords among those circles would also meet at that same point. In that case, the distance between the centre of the range circle and the intersecting point of the common chords among the range circles is equal to the radius of that range circle, that is, the following relation holds.

$$\frac{XC_i}{XC_j} = \frac{r_i}{r_j}, \; where \; 1 \leq i,j \leq 3 \; and \; i \neq j$$

Now, the intersecting point between the common chords moves toward the centre of the range circle drawn from the TOA measurement r_i away from the MS's position and also the boundary of that range circle if r_i contains smaller NLOS error compared to the other TOA measurements. It implies that if r_i contains smaller NLOS error compared to some other TOA measurement r_j then the following relation holds.

$r_i - XC_i > r_j - XC_j$

Now rearranging the above constraint and normalizing it by XC_j , we get

$$\frac{XC_j - XC_i}{XC_j} > \frac{r_j - r_i}{XC_j} > \frac{r_j - r_i}{r_j} \; (\because r_j > XC_j), \; where \; 1 \leq i,j \leq 3 \; and \; i \neq j$$

Again rearranging the above constraint and subtracting 1 from both sides, we get

$$\frac{XC_i}{XC_j} < \frac{r_i}{r_j}$$

$$\Rightarrow \frac{XC_i}{r_i} < \frac{XC_j}{r_j}$$

Therefore, TOA measurement r_i incorporates least amount of NLOS error, if the value of corresponding XC_i/r_i is least.

3.4 Formulation of MGALE Scheme for Estimating the Position of the MS

The objective of the proposed MGALE algorithm is to confine the MS's location estimate within the region bounded by the intersecting point of the common chords among the range circles and two vertices of $\triangle ABC$ that lie on the boundary of the range circle associated with TOA measurement having small-est amount of NLOS error by adopting the geometric constraints between the position of BSs and the MS into the formulation of the 2D GALE scheme.

The following mathematical formulae are used to determine the intersecting point of the common chords among the circles associated with TOA measurements, values of two parameters γ, γ_e and the expected position of the MS.

The equation of the circle with radius r_i and centered at the position of ith BS (α_i, β_i) is given by

$$(x - \alpha_i)^2 + (y - \beta_i)^2 = r_i^2 \tag{7}$$

The intersecting point P_{int} of the common chords among the circles centered at $(\alpha_1, \beta_1), (\alpha_2, \beta_2), (\alpha_3, \beta_3)$ and with radius r_1, r_2, r_3 respectively can be obtained by the following equation.

$$P_{int} = A^{-1}B, \tag{8}$$

where

$$A = \begin{bmatrix} 2\alpha_2 - 2\alpha_1 & 2\beta_2 - 2\beta_1 \\ 2\alpha_1 - 2\alpha_3 & 2\beta_1 - 2\beta_3 \end{bmatrix}, \ B = \begin{bmatrix} r_1^2 - r_2^2 - K_1 + K_2 \\ r_3^2 - r_1^2 - K_3 + K_1 \end{bmatrix}$$

and $K_i = \alpha_i^2 + \beta_i^2$, for $i = 1, 2, 3$.

The virtual distance γ and expected virtual distance γ_e are obtained from equations (1) and (2) respectively by redefining P_k as the intersecting point around the subset of the overlap region among the circles. Here, $P_k = P_1(x_1, y_1)$, $P_2(x_2, y_2)$ and $P_3(x_3, y_3)$ and these are denoted by points A1, B and C respectively when TOA measurements from three BSs are r_1, r_2 and r_3 as shown in fig. 2. To confine the expected position of the MS within the sub region bounded by the intersecting point of the common chords among the range circles and two vertices of $\triangle ABC$ that lie on the boundary of the range circle corresponding to least amount of NLOS error, the proposed MGALE algorithm selects the proper value for the weighting coefficient w_k with respect to point P_k around that bounded region, where $k = 1, 2, 3$. The selection of values for weighting coefficients w_1, w_2 and w_3 considers the geometry of the range circles rather than taking into account the standard deviation of the TOA measurements and these are computed based on the following argument.

If r_j contains larger NLOS error compared to some other TOA measurement r_i, the MS's position should be nearer to the intersecting point around the overlap region that lies inside the range circle associated with r_j. It is away from the boundary of the circle with radius r_j and also the intersecting point that lies within the circle associated with r_i (where $1 \le i, j \le 3$ and $i \ne j$). Based on the above argument the weighting coefficient w_j is defined as

$$w_j = \frac{XC_j/r_j}{\sum_{i=1}^{m}(XC_i/r_i)} \tag{9}$$

The coordinates of expected position of the MS, denoted by $P_e = (x_e, y_e)$, are obtained by the following equation.

$$x_e = \sum_{k=1}^{m} w_k x_k, \ y_e = \sum_{k=1}^{m} w_k y_k \qquad (10)$$

To confine the estimeted MS's position within the sub region $\triangle A1BC$ when three TOA measurements are r_1, r_2 and r_3 respectively, a new parameter virtual noise (n_{r_k}) is introduced.

Definition 1 (Virtual Noise)
The parameter n_{r_k} represents the amount of virtual noise incorporated into TOA measurement r_k with respect to the MS's expected position where $k = 1, 2, 3$ and it is computed by the following equation.

$$n_{r_k} = \frac{r_k - \sqrt{(\alpha_k - x_e)^2 + (\beta_k - y_e)^2}}{c}, \qquad (11)$$

where (α_k, β_k) is the positional coordinates of the kth BS and c is the speed of light.

Based on the 2D GALE algorithm associated with three TOA measurements[16], the MS's position can be estimated within two computing iterations. The intermediate location estimate z' after the first step of the 2D GALE scheme is obtained as follows.

$$z' = \begin{bmatrix} x'_i \ y'_i \ R' \end{bmatrix}^T = (G^T \Psi^{-1} G)^{-1} G^T \Psi^{-1} F, \qquad (12)$$

where (x'_i, y'_i) represent the intermediate location estimation of the MS and $R' = x'^2_i + y'^2_i$. The matrices G, F and Ψ are obtained as follows.

$$G = \begin{bmatrix} -2\alpha_1 & -2\beta_1 & 1 \\ -2\alpha_2 & -2\beta_2 & 1 \\ -2\alpha_3 & -2\beta_3 & 1 \\ -2\alpha_\gamma & -2\beta_\gamma & 1 \end{bmatrix}, \ F = \begin{bmatrix} r_1^2 - K_1 \\ r_2^2 - K_2 \\ r_3^2 - K_3 \\ \gamma_e^2 - K_\gamma \end{bmatrix}, \ \Psi = 4c^2 BQB,$$

where $K_j = \alpha_j^2 + \beta_j^2$ $\alpha_\gamma = \frac{1}{3}(x_1 + x_2 + x_3)$, $\beta_\gamma = \frac{1}{3}(y_1 + y_2 + y_3)$,
$\quad k_\gamma = \frac{1}{3}(x_1^2 + x_2^2 + x_3^2 + y_1^2 + y_2^2 + y_3^2)$, $B = diag\{r_1, r_2, r_3, \gamma_e\}$,
$Q = diag\{n_{r_1}^2, n_{r_2}^2, n_{r_3}^2, \sigma_{\gamma_e}^2/c^2\}$.
Here n_{r_i} denotes the virtual noise incorporated into range measurement r_i for $i = 1, 2, 3$ and σ_{γ_e} corresponds to the standard deviation of γ_e.

The final location of the MS after the second step of 2D GALE scheme is obtained as follows

$$p' = [x' \ y']^T = \left[(G'^T \Psi'^{-1} G')^{-1} G'^T \Psi'^{-1} F' \right]^{\frac{1}{2}}, \qquad (13)$$

where $G' = \begin{bmatrix} 1 & 0 & 1 \\ 0 & 1 & 1 \end{bmatrix}^T$, $F' = \begin{bmatrix} x'^2_i \ y'^2_i \ R' \end{bmatrix}^T$, $\Psi' = 4B'(G^T \Psi^{-1} G)^{-1} B'$,
$B' = diag\{x'^2_i, \ y'^2_i, \ 1/2\}$.

3.5 Overview of MGALE Algorithm

The block diagram describing the different steps of the proposed MGALE algorithm for estimating the 2D location of the MS is given below.

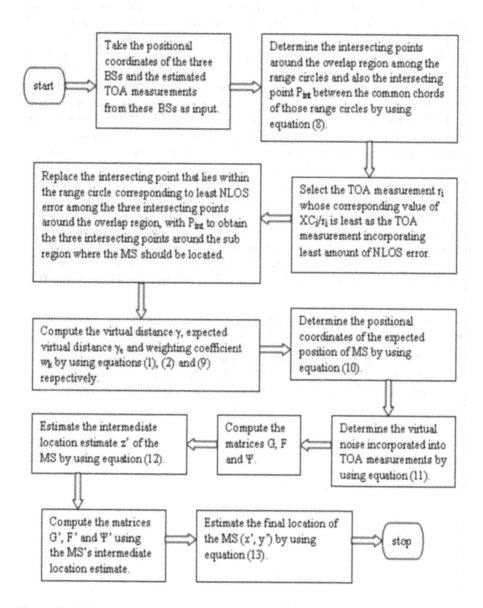

Fig. 4. Block diagram of proposed MGALE algorithm for estimating the 2D location of the MS

4   Performance Evaluations

Simulations were performed to corroborate our proposed location estimation algorithm and to compare the 2D localization accuracy provided by our proposed scheme to that provided by GALE scheme. We have considered the same noise models described in [16] based on extensive field experiments and observations. The measurement noise of TOA signal is assumed to be Gaussian Random variable with zero mean and standard deviation $\sigma_m = 10m$, whereas NLOS error incorporated into TOA measurement is assumed to be non-negative random variable. Three BSs are assumed 2400 meters apart. The home BS is located at $(0,0)$ in meters, whereas the other two BSs are located at $(1200, 1200\sqrt{3})$ and $(-1200, 1200\sqrt{3})$ respectively. The MS is assumed to be located at $(800 \cdot \epsilon - 400, 400\sqrt{3})$ in meters, where ϵ represents a uniformly distributed random number in the range $[0, 1]$, to make the MS moving around the home BS along the line $y = 400\sqrt{3}$.

Table 1 shows the location estimation error acquired by GALE scheme and MGALE scheme under different percentage of position error with $\tau_m = 0.3\mu s$. The estimation error of MS's position is determined by $\Delta p' = \|p' - p^0\|$, where p^0 is position of the MS and p' is the final location estimate from the location estimation algorithm. Table 1 show that our proposed MGALE scheme achieves far better location estimation accuracy compared to GALE scheme in the presence of NLOS errors. This can be explained as follows. MGALE scheme tries

Table 1. Performance Comparison between the Location Estimation Algorithms for the 2D Case with Three TOA Measurements (Estimation Error (m))

	10 %	20 %	30 %	40 %	50 %	60 %	70 %	80 %	90 %	100 %
GALE	29.5	71.3	88.4	113.2	143.6	164.4	188.4	275	313.6	482.1
MGALE	14	9	21.9	42.9	71.3	92	116.8	210.9	254.3	445.1

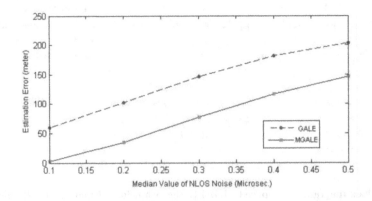

Fig. 5. GALE vs MGALE with 50% of position error under different values of τ_m

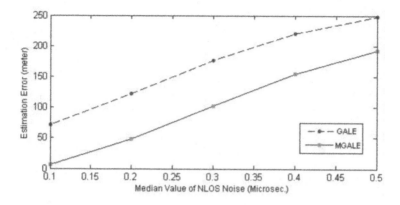

Fig. 6. GALE vs MGALE with 67% of position error under different values of τ_m

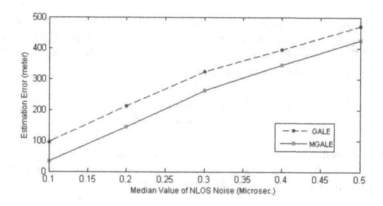

Fig. 7. GALE vs MGALE with 95% of position error under different values of τ_m

to confine the MS's expected position within a smaller region compared to that used by GALE scheme.

Then it attempts to estimate the MS's position by considering the difference between the distance of the MS's expected position from a BS and the TOA measurement corresponding to that BS as the incorporated noise into that TOA measurement rather than the standard deviation of that TOA measurement. We have demonstrated the comparison between the GALE and MGALE scheme under the existence of different values of NLOS noises with 50%, 67% and 95% of position error by fig. 5, 6 and 7 respectively. The different values of NLOS noises are generated by varying the values of τ_m in the range $0.1 \leq \tau_m \leq 0.5\mu s$. Fig. 5, 6 and 7 shows that MGALE scheme achieves better accuracy in location estimation compared to GALE scheme. Fig. 6 and 7 also shows that MGALE scheme meets the required location accuracy defined by FCC for 67% of position

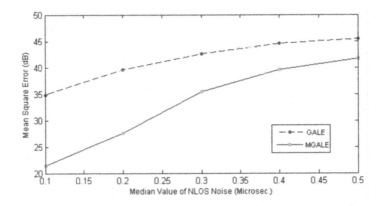

Fig. 8. Performance comparison between GALE and MGALE scheme under 50% of position error: MSE versus median value of NLOS noise

error and 95% of position error respectively with $\tau_m \leq 0.3\mu s$. Fig. 8 shows the Mean Square Error (MSE) of the location estimation using GALE scheme and MGALE scheme with variation in τ_m values. The MSE in decibel (dB) is estimated by the following equation.

$$MSE = \left[10.log\frac{1}{N}\sum_{1}^{N}\left\|p' - p^0\right\|^2\right],$$

where N represents the number of independent runs in the simulations. Fig. 8 also shows that MGALE scheme produces smaller MSE compared to GALE scheme in presence of NLOS errors.

5   Conclusion and Future Work

A modification over existing 2D GALE algorithm with three TOA measurements is proposed in this paper. The proposed MGALE scheme adopts a geometric constraint on the set of possible locations within which the MS can be located into the formulation of 2D GALE scheme to acquire computational efficiency. Moreover, it considers the intersecting point of the common chords of the range circles and also the geometry of those range circles without relying on the standard deviation of TOA measurement to estimate the location of the MS. As the MGALE scheme reduces the area of the region within which MS should be located, it provides better location accuracy compared the GALE scheme under the presence of NLOS error and it can achieve the location accuracy defined by FCC for phase 2. It is possible to utilize the proposed modification for the 2D GALE scheme over the existing 3D GALE scheme to obtain more accurate results for 3D location estimation also. We intend to take this up in the near future.

Acknowledgement

The authors gratefully acknowledge the facilities and support provided by the Director and all other staff members of the School of Mobile Computing and Communication, Jadavpur University, a Centre of Excellence set up under the "University with a potential for Excellence" Scheme of the UGC.

References

1. Revision of the Commissions Rules to Insure Compatibility with Enhanced 911 Emergency Calling Systems. Federal Communications Commission (1996)
2. Rantalainen, T.: Mobile Station Emergency Location in GSM. In: Proc. IEEE International Conference Personal Wireless Communication (PWC 1996), New Delhi, India, February 1996, pp. 232–238 (1996)
3. Chan, Y., Ho, K.: A simple and efficient estimator for hyperbolic location. IEEE Trans. Signal Processing 42(8), 1905–1915 (1994)
4. Friedlander, B.: A passive localization algorithm and its accuracy analysis. IEEE J. Oceanic Eng. OE-12, 234–244 (1987)
5. Caffery Jr., J.: A New Approach to the Geometry of TOA Location. In: Proc. IEEE Vehicular Technology Conference (VTC 2000-Fall), September 2000, vol. 4, pp. 1943–1949 (2000)
6. Cong, L., Zhuang, W.: Hybrid TDOA/AOA Mobile User Location for Wideband CDMA Cellular Systems. IEEE Trans. Wireless Comm. 1, 439–447 (2002)
7. Wylie, M.P., Holtzman, J.: The Non-Line of Sight Problem in Mobile Location Estimation. In: Proc. Fifth IEEE International Conference Universal Personal Communication (IUPC 1996), vol. 2, pp. 827–831 (1996)
8. Chen, P.-C.: A nonline-of-sight error mitigation algorithm in location estimation. In: Proc. IEEE Wireless Communication and Networking Conference (WCNC 1999), pp. 316–320 (1999)
9. Cong, L., Zhang, W.: Non-line-of-sight error mitigation in TDOA mobile location. In: Proc. IEEE Global Telecommunications Conference (GLOBECOM 2001), vol. 1, pp. 680–684 (2001)
10. Xiong, L.: A selective model to suppress NLOS signals in angle-of-arrival (AOA) location estimation. In: Proc. IEEE International Symposium on Personal, Indoor and Mobile Radio Communications (PIMRC 1998), Boston, MA, vol. 1, pp. 461–465 (1998)
11. Wang, X., Wang, Z., O'Dea, B.: A TOA-Based Location Algorithm Reducing the Errors Due to Non-Line-of-Sight (NLOS) Propagation. IEEE Trans. Vehicular Technology 52, 112–116 (2003)
12. Al-Jazzar, S., Caffery, J.: ML and Bayesian TOA location estimators for NLOS environments. In: Proc. IEEE Vehicular Technology Conference (VTC 2002), pp. 1178–1181 (2002)
13. Venkatraman, S., Caffery Jr., J., You, H.R.: A Novel ToA Location Algorithm Using LoS Range Estimation for NLoS Environments. IEEE Trans. Vehicular Technology 53(5), 1515–1524 (2004)
14. Venkatraman, S., Caffery Jr., J.: Hybrid TOA/AOA Techniques for Mobile Location in Non-Line-of-Sight Environments. In: Proc. IEEE Wireless Communication and Networking Conference (WCNC 2004), March 2004, vol. 1(6), pp. 274–278 (2004)

15. Liao, J.F., Chen, B.S.: Robust Mobile Location Estimator with NLOS Mitigation Using Interacting Multiple Model Algorithm. IEEE Trans. Wireless Comm. 5(11), 3002–3006 (2006)
16. Feng, K.-T., Chen, C.-L., Chen, C.-H.: GALE: An Enhanced Geometry-Assisted Location Estimation Algorithm for NLOS Environments. IEEE Trans. Mobile Computing 7(2), 199–213 (2008)
17. Al-Jazzar, S., Caffery, J., You, H.: A scattering model based approach to NLOS mitigation in TOA location systems. In: Proc. IEEE Vehicular Technology Conference (VTC 2002), pp. 861–865 (2002)
18. Greenstein, L.J., Erceg, V., Yeh, Y.S., Clark, M.V.: A New Path-Gain/Delay-Spread Propagation Model for Digital Cellular Channels. IEEE Trans. Vehicular Technology 46, 477–485 (1997)
19. Jeong, Y., You, H., Lee, C.: Calibration of NLOS error for positioning systems. In: Proc. IEEE Vehicular Technology Conference (VTC 2001), vol. 4, pp. 2605–2608 (2001)

Predicting User-Cell Association in Cellular Networks from Tracked Data

Kateřina Dufková[1], Jean-Yves Le Boudec[2], Lukáš Kencl[1], and Milan Bjelica[3]

[1] R&D Centre for Mobile Applications (RDC), Czech Technical University in Prague
Technicka 2, 166 27 Prague 6, Czech Republic
{katerina.dufkova,lukas.kencl}@rdc.cz
[2] Ecole Polytechnique Fédérale de Lausanne
EPFL, CH-1015 Lausanne, Switzerland
jean-yves.leboudec@epfl.ch
[3] Faculty of Electrical Engineering (ETF), University of Belgrade
Bulevar kralja Aleksandra 73, 11120 Belgrade, Serbia
milan@etf.rs

Abstract. We consider the problem of predicting user location in the form of user-cell association in a cellular wireless network. This is motivated by resource optimization, for example switching base transceiver stations on or off to save on network energy consumption. We use GSM traces obtained from an operator, and compare several prediction methods. First, we find that, on our trace data, user cell sector association can be correctly predicted in ca. 80% of the cases. Second, we propose a new method, called "MARPL", which uses Market Basket Analysis to separate patterns where prediction by partial match (PPM) works well from those where repetition of the last known location (LAST) is best. Third, we propose that for network resource optimization, predicting the aggregate location of a user ensemble may be of more interest than separate predictions for all users; this motivates us to develop soft prediction methods, where the prediction is a spatial probability distribution rather than the most likely location. Last, we compare soft predictions methods to a classical time and space analysis (ISTAR). In terms of relative mean square error, MARPL with soft prediction and ISTAR perform better than all other methods, with a slight advantage to MARPL (but the numerical complexity of MARPL is much less than ISTAR).

1 Introduction

Prediction of future user location is useful to a number of applications, including home automation, road traffic management, wearable computers and context aware applications [1,2,3,4]. We are interested in applying location prediction to wireless cellular networks (GSM networks). We seek to estimate the future number of users in different parts of the network, with granularity of a Base Transceiver Station (BTS).

This may have many applications, such as economizing the rental cost of virtual networks, crowd management, provision of real-time network services, or

R. Fuller and X.D. Koutsoukos (Eds.): MELT 2009, LNCS 5801, pp. 19–33, 2009.
© Springer-Verlag Berlin Heidelberg 2009

reduction of energy consumption. For example, it is shown in [5,6] that turning off some of the BTSs when there are few users to serve, and associating these users to neighbouring cells, leads to significant energy savings while maintaining quality of service. Indeed, telephony network operators identify scaling of energy needs with traffic through sleep mechanisms as one of the research challenges of interest for them [7].

As a first step, we would like to evaluate whether it is possible to make some predictions of user association with BTSs, and which prediction methods can be of help. The time scale is 2min, motivated by typical deployment times for near real time network management. Our approach is based on mining the User-Cell association records obtained by active tracking [8]. We evaluate several prediction methods, such as Prediction by Partial Match (PPM), which was successfully used in [1] for location prediction of single users and LAST, which takes as prediction the last visited location. The results motivate us to propose a new method, called "MARPL", which uses Market Basket Analysis to separate patterns where PPM works well from those where LAST is best.

Next, we argue that, in our context, one should make a distinction between *hard* and *soft* prediction. The former predicts the most likely location, whereas the latter gives a spatial distribution. We show how one can transform the hard prediction methods of interest into soft prediction methods. We find that soft predictions are more accurate on our data when tracking an ensemble of users. As a benchmark, we also compare to a classical time and space analysis (ISTAR). The main contributions of the paper are:

- description of a hard prediction method that builds on PPM and Market Basket Analysis to improve prediction;
- transformation of a hard prediction method into a soft prediction method, better suited to the prediction of total number of users at a location;
- comparison, using operator data, of PPM, MARPL, LAST and ISTAR;
- conclusion that user cell sector association can be correctly predicted in ca. 80% of the cases. In term of relative mean square error of user ensemble location estimation, soft methods are better than hard ones, and MARPL with soft prediction and ISTAR perform better than PPM or LAST, with a slight advantage to MARPL (with the added benefit of lower numerical complexity).

The rest of the paper is organized as follows. Section 2 describes the state of the art. Section 3 describes our experimental data. In Section 4 we describe the prediction methods we use. Section 5 presents experimental results and Section 6 concludes the paper.

2   Related Work

Location is an important feature for many applications, and wireless networks can better serve their clients by anticipating client mobility.

González *et al.* in [9] study the trajectories of 100000 mobile phone users over a six-month period. They conclude that the individual travel patterns collapse into a single spatial probability distribution, indicating that it is possible to obtain the likelihood of finding a user in a given location. This further implies that it is possible to quantify the general phenomena driven by human mobility.

Some authors investigate how to obtain datasets which could reliably represent the user's mobility patterns. Sohn *et al.* in [10] showed how coarse-grained GSM data from mobile phones (e.g. readings like signal strength, cell IDs and channel numbers of nearby base station towers) could be used to recognize high-level properties of user mobility. Ashbrook and Starner showed how locations of significance could be automatically learned from GPS data at multiple scales [3]. They describe a system that clusters these data and incorporates them into a predictive Markov model of user's movements. The potential applications of such models would include both single and multi user scenarios. Zang and Bolot in [11] mine more than 300 million call records from a large cellular network operator to characterize user mobility and create mobility profiles. They use passive network monitoring namely in the form of *Per Call Measurement Data* (PCMD) analysis. PCMD records contain data about voice, SMS and data calls performed in the network together with the initial and final cell that served the call. The authors focus mainly on cells where users make call, while we focus purely on user mobility (our data set does not even contain information about calls). Contrary to all these approaches, we use a data set obtained by *active tracking* of selected users' cell associations, without any further "external" location indicators (such as GPS).

Another group of papers investigates methods for predicting user's location. Song *et al.* in [12] present extensive evaluation of location predictors, using a two-year trace of over 6000 users of a Wi-Fi campus network. Even the simplest classical predictors could obtain median prediction accuracy of about 72% over all users with sufficiently long location histories, although accuracy varied widely from user to user. The simple Markov predictors performed comparably or better than the more complicated LZ predictors, with smaller data structures.

There exists a close relation between prediction of discrete sequences and lossless compression algorithms. Begleiter *et al.* in [13] studied the performance of a number of prominent algorithms for prediction of discrete sequences over a finite alphabet, using variable order Markov models. The results show that *Prediction-by-Partial-Match* (PPM) algorithm performed the best. In this paper, we use their implementation of the so-called PPM–C method.

In [1], Burbey and Martin applied the PPM algorithm to data including both temporal and location information. Tests on data traces from IEEE 802.11 wireless network showed that a first-order PPM model had 90% success rate in predicting the user's location, while the third order model was correct 92% of the time. However the studies [12,13,1] were performed on data with different attributes, and an order-of-magnitude lower number of distinct locations, or general states, than in our study.

In this work, we discuss using probabilistic (soft) and aggregate predictions for tracking an ensemble of users. When forecasting the aggregate of variables measured over time and in different regions, it is plausible to assume that the individual components will be spatially correlated. Giacomini and Granger investigate forecasting of a Space-Time Autoregressive model aggregate [14]. Min *et al.* further exploit spatio-temporal correlations to road traffic prediction [4]. Their approach inspired us to formulate the *Integrated Space-Time Auto Regressive* prediction model (ISTAR) (see Section 4.3).

Amongst other papers, Hightower and Borriello used a probabilistic approximation algorithm implementing a Bayes filter, known as *particle filter*, to estimate location [15]. Like us, they also use spatial probability distributions, but they focus rather on indoor localization with an order of magnitude higher precision. Thus, their work is not directly applicable to our dataset. Bauer and Deru notice that relevance of some piece of information is connected to the places a user is likely to visit [16]. They used a variety of machine-learning techniques to derive motion profiles of WLAN users. Their primary goal was not location prediction; instead, they use these profiles to recommend the information which might become useful to the observed user in the foreseeable future.

We end this section with a brief overview of traffic prediction models for wireless networks. Shu *et al.* used seasonal autoregressive integrated moving average (ARIMA) model to capture the behavior of a GSM network traffic stream [17]. Tikunov and Nishimura use a technique known as Holt-Winter's exponential smoothing [18], while Hu and Wu use chaos theory [19].

3   Experimental Data

Mobile cellular networks contain various user data that can be used for location estimation. In the spatial domain, typically the granularity is the *user-cell association*. Finer precision may be gained using triangulation from multiple base stations, but this requires additional sophistication (such as location services platforms), either on the user terminal or on the network side.

Call Detail Records (CDRs) are stored by the telephony network operators. They contain traffic data, including cell association, but only of active users. Mobile terminals themselves may also report their GPS coordinates or currently visible cells (e.g. Google Latitude [20]) over the network, but this requires user cooperation. Cell association of passive, non-communicating users, is beyond the reach of majority of methods, as those users are reporting their location only sporadically using a procedure called location update. A location update is done when a user crosses boundaries of the so called "location areas" (those are geographically large, consisting of hundreds of cells) or after a significant time (order of hours for the network studied in this work). Thus passive users must be tracked actively — the user-cell association observations have to be polled or user-reported.

Data used in this work were obtained by *active tracking of a group of mobile phone users* (unlike in [9]), using the platform from [8]. The platform allows to periodically poll and store cell association of a set of users in a real-time manner and without user cooperation. The users were selected from a list of users who did a location update in the studied network recently, the focus group being foreign roamers. The polling interval was set to 2 minutes, and the association was recorded for all selected users, including passive. The trace contains 72 hours of tracking in December 2008, with 2731 distinct real users moving around in an existing country-wide GSM network. The total number of cells visited by the users was 7332 (not all cells of the network were visited).

For each user, we obtain a sequence of his/her associations, each being either a *Cell identification*, or one of the special states: *Offline*, for users having switched their mobile phones off; *Rival*, for users having left the network to a rival national mobile operator; and *Abroad*, for users having left the network to a foreign operator. The state space thus contains 7335 states.

The trace of user-cell associations represents a sequence of regular location observations [12], the spatial dimension of user mobility. Although previous work has experimented with incorporating both time and space into a single sequence [1], due to the high number of distinct states and amount of data available we chose to deal just with the spatial dimension. Thus, we have removed the *Offline* state from the data, as it seems to depend rather on time of day heavily. We split the user traces around the *Offline* state.

When analyzing user mobility, we observe that the probability of staying at the same location (i.e. being associated with the same cell) is very high and only slowly decreasing over time (see Fig. 1, left), in harmony with [9]. The fashion of selecting users for the tracking implies their higher mobility at the beginning of tracking, as majority of the users are put into the tracking when they are moving, the typical case being a roamer entering country (and the studied network) traveling to a particular destination (see Fig. 1, right).

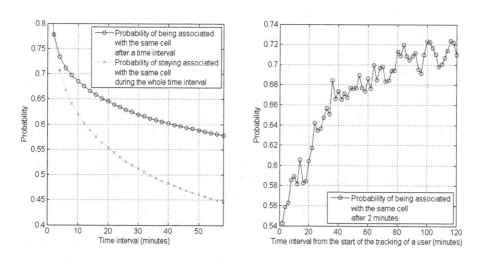

Fig. 1. Left: Probability of a user being associated with the same cell, for different time intervals, mean values over whole tracking. **Right:** 2-minutes mobility of users as function of time interval from the start of the tracking. Due to specific focus on roamers, mobility is higher at the start of the tracking.

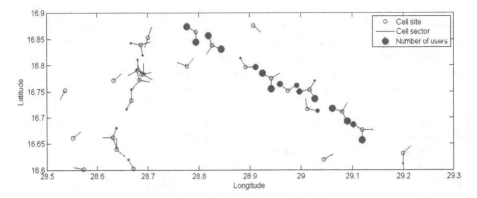

Fig. 2. Example of a cellular network with BTS sites hosting multiple cells with different transceiver directions. A road is recognizable from the higher numbers of users.

4   Predicting Location

Assume we have a finite set of users $\mathbb{I} = \{1, 2, ..., I\}$ and a finite set of cells (base stations, access points, etc.) $\mathbb{J} = \{1, 2, ..., J\}$ of a cellular network. Assume we can observe the cell association $a^i(t) \in \mathbb{J}$ for any user $i \in \mathbb{I}$ and any time $t \in \mathbb{N}$. Let $A^i = \{a^i(t)\}$, $t \in \mathbb{N}$ be a sequence of observations of cell association for a user $i \in \mathbb{I}$ over discrete equidistant time slots. Let $Y_j(t)$ be the number of users associated with cell $j \in \mathbb{J}$ in a time slot $t \in \mathbb{N}$.

4.1   Hard vs. Soft Decisions

Assume H^i is a sequence of previous associations of a user i. We define the *hard decision location prediction problem* as the task of finding a single location $j \in \mathbb{J}$ with the highest $Prob(j|H^i)$, where the user i will most likely be at the next time slot. We define the *soft decision location prediction problem* as the task of constructing a vector $U^i = [u^i_j]$, $u^i_j = Prob(j|H^i)$ of probabilities for a user $i \in \mathbb{I}$ to be at any possible location $j \in \mathbb{J}$.

While the predictors that provide a hard decision on the next location of the user are useful in many applications, the "winner takes all" strategy does not have to be optimal for all applications. One of them is the application we study in this paper, where *aggregation* is used to obtain network-wide statistics about numbers of users associated with individual cells (see Figure 2).

We formalize the task as follows: Knowing $a^i(s)$, $i \in \mathbb{I}$, $s \in \{1, 2, ..., t-1\}$ and $Y_j(s)$, $j \in \mathbb{J}$, $s \in \{1, 2, ..., t-1\}$ we want to predict $Y_j(t)$. For practical reasons, as we do not want to store much historical data, we want to base the prediction just on the last r values, i. e. on the values related to $s \in \{t-r, t-r+1, ..., t-1\}$.

4.2   Individual Hard Decision Methods

LAST predictor. The simplest possible predictor, which always uses the last known value as the prediction, will be used as a reference for proposed methods.

PPM predictor. The well-known *Prediction-by-Partial-Match* (PPM) algorithm that uses variable order Markov models. We use implementation of the so-called PPM–C method provided by [13].

MARPL predictor. We propose a method called MARPL (MARket basket analysis + Ppm + Last), which combines PPM and LAST predictors, after splitting the problem into subproblems according to the last few associations of the user, and choosing the best strategy for every subproblem independently.

The splitting is loosely inspired by the Market Basket Analysis method [21] and its way of discovering hidden rules in the data, with the difference that the original method was intended for unordered sets of elements instead of ordered sequences. We construct set of all possible rules of order r, each rule being of the form $H_1 H_2...H_r \to P$, where $H_s \in \{A, B, C...\}$ represents the history of the last r associations of a user and $P \in \{A, B, C...\}$ represents the predicted association. The $A, B, C, ...$ symbols are wildcards as we are interested in generally applicable rules. For example rule $AABB \to B$ represents the situations where, after observing a cell A twice and then another cell B twice, the next cell is B.

We define applicability and reliability of a rule as follows (L(rule) denotes the left side of a rule, R(rule) the right side of a rule):

$$\text{Applicability(rule)} = \frac{\#\text{ possible usages}}{\#\text{ all predictions}} = Prob\left(L(\text{rule})\right), \tag{1}$$

$$\text{Reliability(rule)} = \frac{\#\text{ successful usages}}{\#\text{ possible usages}} = Prob\left(R(\text{rule})|L(\text{rule})\right). \tag{2}$$

We split the problem as follows:

1. Use the LAST predictor on subproblems, where the rule corresponding to the LAST predictor has strictly higher reliability than the PPM predictor success rate (54,5%, see Section 5). See Table 1.
2. Otherwise use the PPM predictor with a fallback to the LAST predictor on cases where the PPM is "not sure". The level of certainty of the PPM prediction can be obtained as the likelihood $Prob$(Predicted symbol$|H^i$); we accept the PPM prediction only if its likelihood is above certain threshold.

Table 1 summarizes results of the analysis for our data and $r = 4$, which proved best in the experiments. The thresholds were set according to the reliability of the LAST predictor on the subproblem (see Table 1). The lower the percentage of good predictions that LAST predictor would make, the lower the threshold and, consequently, the lower the number of fallbacks to the LAST predictor.

The reason we chose to use directly the LAST predictor on some subproblems (instead of using high threshold) is performance. The subproblems where we use LAST predictor together make 77% of the cases, so the MARPL achieves remarkable speedup of the prediction process, compared to the PPM predictor.

Finally, selection of the training data needs care. The staying pattern (rule $AAAA \to A$) is dominant in the dataset, but useless for the PPM predictor, as it will never be used on this kind of data. We considered three training phase strategies — using *all available data*, using *selected overlapping subsequences of length* $r + 1$, and using *selected non-overlapping subsequences of variable length*. The overlapping sequences strategy omitted the sequences that contained just one symbol, the non-overlapping sequences strategy continued to grow the current subsequence until the staying pattern was recognized, and then started a new sequence, omitting the repeating symbols. The selected non-overlapping subsequences proved best in the experiments and will be used further.

4.3   Aggregated Soft Decision Methods

In this section we transform MARPL and PPM predictors to provide *soft decisions*. Then we propose another approach, that does not take into account individual users and predicts the number of users directly.

MARPL soft predictor. The MARPL predictor provided just the single most likely next location. Instead of it a vector $U^i = [u^i_j]$, $u^i_j = Prob(j|H^i)$, $j \in \{1, 2, ..., J\}$ of probabilities for a user i to be at all the possible locations $j \in \{1, 2, ..., J\}$ is now needed. We construct the vector as follows.

- For the subproblems where PPM is used, $u^i_j = Prob(j|H^i)$ where H^i is the association history of user i.
- For the subproblems where LAST is used, $u^i_j = 1$ if j is the prediction obtained by LAST, $u^i_j = 0$ otherwise.

Table 1. Market Basket Analysis for sequences of associations A^i for rules of order 4. Each row represents all rules with the same left side. The rules can be classified into two user behaviour patterns — *stay* and *move*. Staying (represented by the $AAAA \rightarrow A$ rule) prevails greatly, the rest of the rules relate to moving users. The star marks the subproblems where the threshold chosen according to the reliability of the LAST predictor did not perform well, and was changed to more appropriate value.

Rules	Applicability (%)	Reliability (%) A B C D E					LAST reliability	Algorithm	Threshold (%, rounded up)
$A, A, A, A \rightarrow ?$	**66.8**	96	4	-	-	-	96	LAST	-
$A, B, C, D \rightarrow ?$	**8.7**	0	1	2	18	79	18	PPM	18
$A, A, A, B \rightarrow ?$	4.0	29	45	26	-	-	45	PPM	46
$A, B, B, B \rightarrow ?$	4.0	12	70	18	-	-	70	LAST	-
$A, A, B, B \rightarrow ?$	3.0	17	60	23	-	-	60	LAST	-
$A, B, C, C \rightarrow ?$	2.7	3	4	40	53	-	40	PPM	40
$A, A, B, C \rightarrow ?$	2.6	4	6	31	59	-	31	PPM	41*
$A, B, B, C \rightarrow ?$	2.3	4	8	30	58	-	30	PPM	31
$A, A, B, A \rightarrow ?$	1.6	67	21	12	-	-	67	LAST	-
$A, B, A, A \rightarrow ?$	1.6	72	16	12	-	-	72	LAST	-
$A, B, B, A \rightarrow ?$	0.9	55	31	14	-	-	55	PPM	45*
$A, B, A, B \rightarrow ?$	0.7	41	48	10	-	-	48	PPM	49
$A, B, C, B \rightarrow ?$	0.5	8	43	16	32	-	43	PPM	43
$A, B, A, C \rightarrow ?$	0.4	16	8	37	39	-	37	PPM	37
$A, B, C, A \rightarrow ?$	0.3	46	12	17	25	-	46	PPM	46

- By aggregating the vectors U^i for all the users $i = \{1, 2, ..., I\}$ we obtain the prediction $\hat{Y}_j(t) = \sum_{i=\{1,2,...,I\}} u_j^i$.

PPM soft predictor. Created from the PPM predictor by the same procedure as MARPL soft predictor (the second branch is never used).

Integrated Space-Time Auto Regressive model (ISTAR). The proposed method is a time series analysis method inspired by [4] on road traffic prediction. Assume we have an adjacency matrix $A_{i,j}$ such that $A_{i,j} = 1$ if a user can move from location i to location j within one time step (at the highest possible speed). Otherwise $A_{i,j} = 0$. The matrix A is static, derived by comparing the distances between all pairs of BTS with a fixed distance threshold D. Recall that $Y_j(t)$ is the number of users at location j at time t. We apply differencing, as is common in time series analysis, and define $X_j(t) = Y_j(t) - Y_j(t-1)$. The model is:

$$X_j(t) = \sum_{i:A_{i,j}=1} \alpha_{i,j} X_i(t-1) + \beta_j X_j(t-1) + \epsilon(t) \tag{3}$$

where $\epsilon(t)$ is Gaussian white noise. The parameters to be estimated are the matrix α ($J \times J$), the vector β ($J \times 1$) and the noise variance (J is the number of locations). At time t, the prediction for $X_j(t+1)$ is $\hat{X}_j(t) = \sum_{i:A_{i,j}=1} \alpha_{i,j} X_i(t) + \beta_j X_j(t)$. The parameters α and β are estimated by minimizing

$$\hat{\sigma}_t^2 := \frac{1}{tJ} \sum_j \sum_{s=2}^t w^{t-s} \left(X_j(s) - \hat{X}_j(s-1) \right)^2 \tag{4}$$

where w is a "forgetting" factor, close to 1 and less than 1. Finally, the one-step-ahead prediction for $Y_j(t+1)$ is $\hat{Y}_j(t) = \hat{X}_j(t) + Y_j(t)$.

4.4   Algorithm Complexity

The complexity of predicting the next state of the whole network is considered.

PPM & MARPL. Given the implementation we use, the complexity of PPM prediction for I users and histories of r associations is $O(I \cdot J \cdot r^2)$. For MARPL, the complexity of predicting is $O(r)$ for the decision between the PPM and LAST plus $O(1)$ for the 77% of cases where the LAST predictor is used, or PPM prediction complexity for the rest of the cases. For both, the time complexity of learning one sequence of length n is $O(n)$ and the space required for the worst case is $O(r \cdot n)$, where r is the order of the model [13].

ISTAR. Theoretically, the complexity of predicting the next value for all J locations is $O(J^2)$. The complexity of estimating the α and β parameters is determined by the complexity of computing Equation 4 ($O(t \cdot J^2)$ where t is number of time slots) and complexity of minimization. As minimization algorithm we use Matlab function *lsqnonlin* with default Trust-Region-Reflective algorithm (whose complexity is $O(\text{iterations} \cdot \text{parameters})$) on $O(J^2)$ parameters corresponding to the fraction of ones in adjacency matrix A. Thus the overall parameter estimation worst case complexity is $O(t \cdot J^4 \cdot \text{iterations})$ and $O(J^2)$ space is required. Practically, on large networks the matrix A will become sparse and the J^2 factor can be replaced with J^a, $a \in [1, 2)$, leading to $O(t \cdot J^{2a} \cdot \text{iterations})$ complexity.

For our data ($I = 2731$, $J = 7335$, $r = 4$, $t = 60$, $n = $ ca. 170000) the complexity (in terms of both space and time) of soft PPM and soft MARPL is one order of magnitude lower than that of ISTAR.

5   Experimental Results

5.1   Individual Hard Decision Methods

Data. To use the PPM predictor the data need to be divided to training and test groups. The original data of 2731 users were pseudo-randomly split to 20 groups and experiments were repeated 20 times, each time with one group as test data and the rest of groups used as training data. Each test group contained 96681 subsequences of length 4 with correct next association for evaluation purposes.

Comparing MARPL, PPM and LAST predictors. Figure 3 compares the hard predictors by means of both percentage of correct predictions and distribution of distances between the real and predicted cell. Note that 0m distance between the real and predicted cell occurs in two cases — when correct sector on correct base station is predicted (denoted as *OK BTS+sector*), and when another sector on correct base station is predicted (denoted as *OK BTS*). The difference stems from the cellular network architecture, where a base station often holds more transceivers, serving different sectors and cells, most commonly three.

The MARPL predictor performs best, achieving 79.3% success rate when the exact prediction of BTS and sector is required, and 84.4% success rate when the prediction of BTS suffices. From the perspective of predicting user location to switch off under-utilized hardware, the above results are encouraging, as the lower distance errors prevail markedly. We can conclude that the MARPL is able to predict correctly 94% associations with error up to 2500 metres, which is acceptable given the typical cell overlays in cellular networks.

Surprisingly the PPM predictor performs worse than the LAST predictor. The reason is that the LAST predictor builds on the low mobility of users (see Figure 1), while PPM has to deal with problems related to the character of our data — the *high number of distinct cells* to associate with, the consequent *training data shortage* and finally the *PPM predictor behavior when "not sure"*. Here PPM predicts the most frequent symbol of the training data (universal *Rival* state for our data), while having in mind the Figure 1, the best strategy is to predict the last known value. The MARPL predictor overcomes these problems by using PPM on the subset of data coming from moving users, and LAST on the data from staying users.

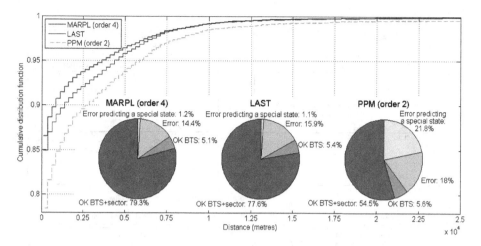

Fig. 3. Comparison of hard predictors, cumulative distribution function of distances between the real and predicted cell. Inset are pie charts showing overall success rate wrt. exact next cell-ID prediction. For both PPM and MARPL only the results of the best performing model are shown for brevity (order 2 for PPM, order 4 for MARPL). PPM is markedly the worst of the predictors, LAST and MARPL provide similar results, with slight advantage of MAPRL. However both PPM and MARPL can be improved by introducing soft decisions, while LAST has no soft decision variant.

5.2   Aggregated Soft Decision Methods

Data. The splitting to training and test data was the same as in previous section. From the test data, just the users with associations history long enough to predict 60 consecutive time slots were selected, which makes 1296 users and total of 77760 predictions in all 60 time slots.

Comparing soft and hard predictors. Fig. 4 compares the aggregated predictions from soft and hard versions of MARPL and PPM predictors by means of mean squared error (MSE) between the vectors $[\hat{Y}_j(T)]$, $j \in \mathbb{J}$ obtained using the predictors, and the real vector $[Y_j(T)]$, $j \in \mathbb{J}$. MARPL consistently achieves lower MSE than the PPM predictor, and soft predictors consistently achieve lower MSE than the hard predictors, both for single group of test users and for all groups. The mean MSE for MARPL soft predictor is 0.070, which is just 66.4 % of the mean MSE of LAST (0.106) and 69.8 % of the mean MSE of PPM soft (0.101).

On our dataset, the growing size of population does not affect the results. While the absolute MSE grows with the number of users in the population, the MSE relative to the number of users remains approximately the same, making the order of the methods stable for all population sizes we considered.

Optimal parameters of the ISTAR model. The parameters of the model are the "forgetting" factor w and distance threshold D, which determines the number

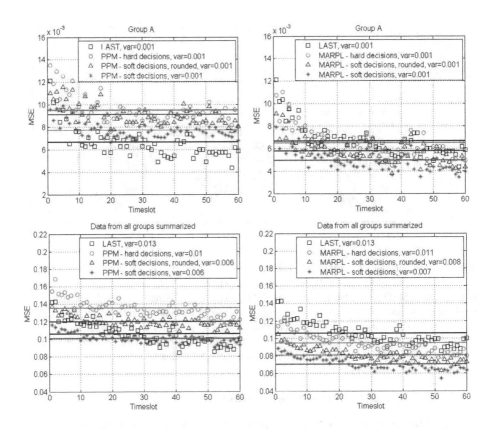

Fig. 4. Comparison of aggregated soft and hard predictors over 60 consecutive time slots, MSE. Top graphs show values for a single group of test data (96 users), bottom graphs for all test groups (1296 users). The left graphs show variants of the MARPL predictor, the right graphs of the PPM predictor. The reason why the MSEs are generally low, especially for single group of test data, is that we have only 96 (or 1296) users moving around 7332 cells, which implies large number of empty cells (where all predictors succeed), pushing the MSE down. Perhaps also surprising is that all predictors improve over time, even though the LAST predictor obviously does not learn from past data. This is due to the diminishing mobility of users over time (see Fig. 1).

of ones in adjacency matrix and thus the computation complexity. Figure 5 concludes that ISTAR improves with higher D and works best for $w = 0.95$.

Comparing aggregated location predictors and ISTAR model. Finally we compare the aggregated results of the location predictors and of ISTAR with optimal parameters. Due to the computational requirements of ISTAR (see Section 4.4), the comparison was feasible on only a subset of 59 cells in one geographical district. The results of location predictors were obtained by restricting the results from the experiment over the entire dataset to the selected cells. This raises the question if it is fair to compare models trained on larger data

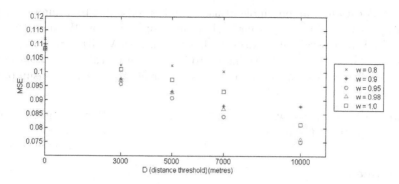

Fig. 5. The ISTAR model performance given by means of MSE for different combinations of parameters. The model improves with higher D and works best for $w = 0.95$.

to ISTAR, but why ignore MARPL's and PPM's capability to train on larger datasets. Regarding test data, the neighborhood errors at the region's borders may influence ISTAR, but not enough users associated to the selected cells for 60 consecutive time slots were available to fairly scale down the location predictors tests. The results (see Table 2) conclude that the MARPL soft predictor performs best out of the studied methods.

Table 2. The overall MSE achieved by the studied methods (ordered from best to worst). We specify the type of results for each method, for real number predictors (soft predictors and ISTAR) rounding is considered.

Method	MARPL	ISTAR	MARPL	ISTAR	MARPL	PPM	LAST	PPM	PPM
Decisions	Soft	-	Soft	-	Hard	Soft	Hard	Soft	Hard
Result	$\mathbb{R}$	$\mathbb{R}$	$\mathbb{R} \to N$	$\mathbb{R} \to N$	N	$\mathbb{R}$	N	$\mathbb{R} \to N$	N
MSE	0.0715	0.0750	0.0864	0.0890	0.0949	0.1228	0.1263	0.1537	0.2144

6 Conclusions

We show that predicting user location within a cellular network in the next time interval, with the granularity of the associated BTS, is a feasible task with acceptable performance. On our experimental data, best results are achieved using a novel prediction method, MARPL, which combines Prediction by partial match (PPM) and LAST location predictor, using Market Basket Analysis. This is an initial result on a limited (size) and specific (roaming clients) data set — general applicability to arbitrary cellular network mobility data will need to be verified in the future.

Further, we argue that the soft, probabilistic prediction methods are more useful in predicting the aggregate location of a user ensemble, as shown using mean square error comparison. Predicting location as a probabilistic vector, or aggregate location of an ensemble of users, makes sense due to a number of potential applications focusing on network resource optimization. We show that the soft methods in general outperform the hard ones, with MARPL requiring fewer resources. In our future work, we intend to focus on practical applications of the predictions for tasks such as economizing cellular network energy consumption.

Acknowledgment. We wish to thank Vodafone Czech Republic a.s. for their generous support of the project.

References

1. Burbey, I., Martin, T.L.: Predicting future locations using prediction-by-partial-match. In: MELT 2008: Proceedings of the first ACM international workshop on mobile entity localization and tracking in GPS-less environments, San Francisco, California, USA, pp. 1–6. ACM, New York (2008)
2. Das, S.K., Cook, D.J., Battacharya, A., Heierman III, E.O., Lin, T.-Y.: The role of prediction algorithms in the MavHome smart home architecture. IEEE Wireless Communications 9, 77–84 (2002)
3. Ashbrook, D., Starner, T.: Using GPS to learn significant locations and predict movement across multiple users. Personal and Ubiquitous Computing 7, 275–286 (2003)
4. Min, W., Wynter, L., Amemiya, Y.: Road traffic prediction with spatio-temporal correlations. In: Proceedings of the Sixth Triennial Symposium on Transportation Analysis, Phuket Island, Thailand (June 2007)
5. Chiaraviglio, L., Ciullo, D., Meo, M., Marsan, M., Torino, I.: Energy-aware UMTS access networks. In: Proceedings of the 11th International Symposium on Wireless Personal Multimedia Communications (WPMC 2008), Lapland, Finland (September 2008)
6. Marsan, M.A., Chiaraviglio, L., Ciullo, D., Meo, M.: Optimal Energy Savings in Cellular Access Networks. In: Proceedings of GreenComm 2009 — First International Workshop on Green Communications, Dresden, Germany (June 2009)
7. Lister, D.: An operator's view on green radio. In: First International Workshop on Green Communications, GreenComm 2009 (June 2009) (keynote presentation)
8. Dufková, K., Ficek, M., Kencl, L., Novák, J., Kouba, J., Gregor, I., Danihelka, J.: Active GSM cell-id tracking: Where did you disappear? In: MELT 2008: Proceedings of the first ACM international workshop on mobile entity localization and tracking in GPS-less environments, San Francisco, California, USA, pp. 7–12. ACM, New York (2008)
9. González, M.C., Hidalgo, C.A., Barabasi, A.L.: Understanding individual human mobility patterns. Nature 453(7196), 779–782 (2008)
10. Sohn, T., Varshavsky, A., Lamarca, A., Chen, M., Choudhury, T., Smith, I., Consolvo, S., Hightower, J., Griswold, W., de Lara, E.: Mobility detection using everyday GSM traces. In: Dourish, P., Friday, A. (eds.) UbiComp 2006. LNCS, vol. 4206, pp. 212–224. Springer, Heidelberg (2006)
11. Zang, H., Bolot, J.C.: Mining call and mobility data to improve paging efficiency in cellular networks. In: MobiCom 2007: Proceedings of the 13th annual ACM international conference on Mobile computing and networking, Montréal, Québec, Canada, pp. 123–134. ACM, New York (2007)

12. Song, L., Kotz, D., Jain, R., He, X.: Evaluating location predictors with extensive Wi-Fi mobility data. In: Proceedings of the 23rd Annual Joint Conference of the IEEE Computer and Communications Societies (INFOCOM 2004), Hong Kong, China, March 2004, vol. 2, pp. 1414–1424 (2004)

13. Ronbeg, R.B., Yona, G.: On prediction using variable order Markov models. Journal of Artificial Intelligence Research 22, 385–421 (2004)

14. Giacomini, R., Granger, C.W.: Aggregation of space-time processes. Boston College Working Papers in Economics 582 (June 2002)

15. Hightower, J., Borriello, G.: Particle filters for location estimation in ubiquitous computing: A case study. In: Davies, N., Mynatt, E.D., Siio, I. (eds.) UbiComp 2004. LNCS, vol. 3205, pp. 88–106. Springer, Heidelberg (2004)

16. Bauer, M., Deru, M.: Motion-based adaptation of information services for mobile users. In: Ardissono, L., Brna, P., Mitrović, A. (eds.) UM 2005. LNCS (LNAI), vol. 3538, pp. 271–276. Springer, Heidelberg (2005)

17. Shu, Y., Yu, M., Liu, J., Yang, O.: Wireless traffic modeling and prediction using seasonal ARIMA models. In: Proceedings of the IEEE International Conference on Communications, 2003 (ICC 2003), Anchorage, Alaska, USA, May 2003, vol. 3, pp. 1675–1679 (2003)

18. Tikunov, D., Nishimura, T.: Traffic prediction for mobile network using Holt-Winter's exponential smoothing. In: Proceedings of the 15th International Conference on Software, Telecommunications and Computer Networks (SoftCOM 2007), Portsmouth, UK, September 2007, pp. 1–5 (2007)

19. Hu, X., Wu, J.: Traffic forecasting based on chaos analysis in GSM communication network. In: Proceedings of the International Conference on Computational Intelligence and Security Workshops (CISW 2007), Harbin, Heilongjiang, China, December 2007, pp. 829–833 (2007)

20. Google: Latitude project, http://www.google.com/latitude/

21. Agrawal, R., Imieliński, T., Swami, A.: Mining association rules between sets of items in large databases. In: SIGMOD 1993: Proceedings of the 1993 ACM SIGMOD international conference on management of data, Washington, D.C., United States, pp. 207–216. ACM, New York (1993)

Discovering Significant Places from Mobile Phones – A Mass Market Solution

Guang Yang

Nokia Research Center
955 Page Mill Road
Palo Alto, CA 94304, USA
guang.g.yang@nokia.com

Abstract. In this paper we propose a mass market solution on mobile phones to discover a user's significant places solely from observed cell IDs. It does not require either cell-ID-to-physical-location mapping or the capability of obtaining multiple cell IDs on the phone simultaneously, and is able to run on virtually any mobile phone today. Our solution is centered around a cell ID clustering algorithm based on temporal correlations. It is able to prevent over-clustering and handles missing data well. We evaluate the solution with real-life data that the author has collected over a period of eight weeks. Results show that we are able to discover not only places of utter importance, but also certain less frequently recurring places and one-time travel destinations that bear significance in one's life.

1 Introduction

Location-based services are among the hottest and fastest-growing mobile applications today. Many such services aim to utilize a user's physical location by mashing it up with local search, advertising or social network services, etc. The user's location is usually obtained from GPS, cell location databases [14] or Wi-Fi hotspot location databases [16], or a combination.

GPS is the dominant localization technology nowadays, but has several drawbacks that make alternative methods also appealing. GPS needs a clear view of the sky and has poor reception inside buildings or in other obstructed environments. It consumes much energy if used continuously, or suffers prolonged lock-on periods if used on demand. GPS also increases the manufacturing cost and is only available in middle-tier to high-end mobile phones today. Assisted-GPS (A-GPS) [3] has partially solved these problems, but needs additional chipset and network support, and the cost issues remain.

In this paper we argue that for some use cases, a cheaper localization solution may also be a better one. One such use case is location-aware user interface. For instance, application shortcuts may be selected and ordered based on the user's particular usage pattern at a specific locale. Another potential use case is contextual content management. An example is for the mobile phone to automatically tag with location information user-generated content including photographs and

R. Fuller and X.D. Koutsoukos (Eds.): MELT 2009, LNCS 5801, pp. 34–49, 2009.
© Springer-Verlag Berlin Heidelberg 2009

video clips. In either case, absolute and/or high-resolution physical location information (e.g. geo-coordinates) may not be necessary as long as the locale is meaningful and distinguishable to the user herself.

Virtually all mobile phones today, as popularly called cell phones, rely on the cellular network infrastructure for communications. They must be able to identify and connect to a cell, and remain connected in operation. The biggest advantages of cell-ID-based localization are the universal availability – in terms of both the device base and network coverage, and near-zero added costs – in terms of both manufacturing and energy consumption. Even with the downsides such as coarse accuracy and lack of easy cell-ID-to-physical-location mapping – a GSM cell may span well over several km^2, and today's cell location databases are far from open/adequate – we will show in this paper that if handled appropriately, cell IDs alone may still reveal plenty of a user's location patterns. Better yet, it requires no Internet connections; all data collection and computation may be performed on the phone in a privacy-preserving way.

The rest of the paper is organized as follows. Section 2 gives an overview of related mechanisms and their respective difficulties in our target mass market scenario. Section 3 describes two key assumptions that we take in this paper, defines formally the problem to be solved, and lists three types of significant places of our interest. Section 4 presents a naive cell ID clustering algorithm as the very basic foundation of our solution, followed by algorithmic improvements in Section 5 to address several shortcomings. In Section 6 we evaluate our work with eight weeks of real-life data collected on the author's primary cell phone, and finally Section 7 discusses future work and concludes the paper.

2   Related Work

Discovering significant places using GPS is a well explored area in the literature. [1] proposes to cluster GPS coordinates into meaningful locales based on geo-distances. It looks at both spatial and temporal correlations among GPS coordinates in the trace, tuning threshold parameters to find the desirable clustering results. The solution is flexible and may recursively find places at multiple hierarchical levels; its geo-distance-based clustering method, however, cannot be easily generalized to non-GPS technologies, as we will see in Section 3. [11] extracts a user's activities and significant places from GPS traces using hierarchical conditional random fields. Its primary novelties are to take high-level context into consideration when detecting significant places, and the ability to train and apply models across users. The main difficulties of applying it in our scenario are the dependency on maps and fairly intense computation.

Among the GPS-less localization methods, many focus on measuring the received signal strengths from Wi-Fi base stations and/or GSM cell towers to pinpoint a user's physical location [8][10]. These methods typically rely on mapping databases to convert Wi-Fi base station MAC addresses and/or GSM cell IDs to their geo-coordinates. In this paper we do not assume availability of such mapping databases. Additionally, Wi-Fi-based localization has disadvantages in network coverage and device penetration compared to GSM-based methods. GSM-based methods have their own limitations – as we will elaborate later, it is practically difficult to obtain information from multiple GSM cells simultaneously on mainstream mobile phones, due to lack of such support in today's mass market implementations.

Several other interesting projects in GPS-less localization are [2][4][6]. [2] compares Wi-Fi scanning data, including MAC addresses and received signal strengths, on two mobile devices and determines whether they are in proximity according to a Gaussian Mixture Model. It does not require knowledge on the physical locations of scanned Wi-Fi devices. [4] presents a pure cell-ID-based solution for user tracking. It requires support from the network infrastructure and knowledge on the physical locations of cell towers. BeaconPrint [6] moves away from received signal strengths and uses *beacons* to characterize places. It works well with Wi-Fi in which it is easy to obtain information from multiple base stations at the same time; with GSM the aforementioned difficulty in obtaining information from multiple cells remains.

It is worth pointing out that [7], built on top of Place Lab [10], introduces a time-based clustering algorithm that has in part inspired our work in this paper. Unlike standard clustering algorithms, the proposed method does not specify the number of clusters to be created as a parameter, and its light computation favors resource-limited mobile devices. Like other projects based on Place Lab, however, it too relies on a mapping database to obtain geo-distances between data points, and thus does not apply directly to our target problem.

The most directly related work to our paper is perhaps [9], which also proposes to use solely a timestamped sequence of cell IDs to discover a user's significant places. There exist two distinctions in our solution, though. First, the cell ID clustering algorithm is different – [9] is based on length of stay, i.e. statistics, while ours is based on temporal correlation in the input sequence; our cluster-merging algorithm is also more restrictive to prevent over-clustering along roads. Second, [9] applies an aging algorithm to gradually purge non-recurring places, while we see such places as potentially special to the user. We explicitly address non-recurring significant places in this paper. We do not address *routes* in this paper but plan to investigate this problem in detail in the near future.

3   Assumptions and Problem Definition

3.1   Assumptions

We make two key assumptions to the problem to be solved in this paper:

1. There are no cell-ID-to-physical-location mapping databases available.
2. Mobile phones can only obtain information about the cell that it is currently connected to.

Removal of either assumption would certainly make the problem easier to tackle, but the state-of-the-art in the mobile phone industry warrants both when our target is the global mass market. Exact locations of cell towers are well-guarded secrets by mobile carriers and generally unavailable to the public. Commercial location-based services often rely on proprietary databases [12][13], or crowd-sourcing initiatives [14] that are far from being complete. Obtaining more than one cell ID simultaneously is feasible in theory, but few mobile phones support it in practice. In fact, even getting one cell ID involves completely different APIs across manufacturers because programming languages such as Java ME do not provide such a standard API. It is likely that both assumptions will hold in the foreseeable future.

3.2 Problem Definition

The input to our problem is a sequence of N observed cell IDs along with timestamps. Denote cell IDs as C_i and corresponding timestamps as T_i, $i = 0, 1, \ldots,$ $N - 1$, the input is in the form of $\{(C_i, T_i)\}$ where (T_i) is a strictly ascending sequence. Our goal is to find significant places $\{P_j \mid j = 0, 1, \ldots, M - 1\}$ from the input. Similar to [7], the term *place* refers to a user-specific, semantically meaningful locale. It is worth pointing out that our goal here is merely to find places that bear certain degrees of importance to the user; we do not intend to label these places, as labeling is a different (and difficult) problem that deserves its own full attention.

3.3 Defining Significance

We are interested in two dimensions of significance – length of stay and recurrence. To qualify as "significant", a place must score high on at least one dimension; places that neither have a long time of stay nor demonstrate a strong recurring pattern are not of our interest in this paper.

Places that score high on both dimensions are the places of utter importance to the user, e.g. "Home" or "Work". We expect a user to have just a handful of such places, and they may be quite obvious to identify.

Non-recurring places that show a relatively long time of stay in a limited time frame are often business trip or vacation destinations. Over a longer period (such as several months) these places may not seem as important in terms of absolute length of stay, but since they are out of one's daily routine and thus "abnormal" rather than "normal", the user often remembers them as "special occasions" that bear certain significance in life.

The third type, i.e. recurring places without a long time of stay, is more subtle. "Starbucks" and "Child's School" are examples here, where a user may spend only a few minutes on any given day. Recurrence along does not necessarily lead to significance, e.g. a commuter may pass the same on-ramp meter at the freeway entrance every morning, but the ramp barely has any semantic meaning to the user. Later in the paper we will address this with real-life data; however whether/how we can tell "Child's School" from "On-ramp of I-280 at De Anza Blvd" remains a research problem to be further investigated.

4   Clustering Cell IDs into Places

Given the input format in Section 3, a place is represented as a set of IDs of geographically co-located or nearby cells. It is a set rather than a single ID because cells are often deployed with overlapping to enhance connectivity robustness. Even stationary, a mobile phone may dynamically hand off to a different cell if the new cell is considered "better" than the current one. Observed "raw" cell IDs must be clustered first before further steps are taken.

We want to find cells that are co-located or close to each other. Without cell-ID-to-physical-location mapping databases, a clustering algorithm solely on the input is needed. The basic idea is to find minimum circular subsequences of cell IDs in the input. A *circular subsequence* is defined as a subsequence starting and ending with the same cell ID and contains at least two different cell IDs. A *minimum circular subsequence* is a circular subsequence that consists of no circular subsequences in itself.

A minimum circular subsequence indicates that the mobile phone has "returned" to where it was at the beginning. This return may or may not have involved physical movement of the phone. To exclude from the clustering process those cases in which the phone has indeed traveled long distances physically before coming back, we limit our interest in minimum circular subsequences of low cardinalities. The *cardinality* of a sequence is defined as the number of different symbols in it. For example, in "$\ldots XABBBCCAY\ldots$", "$ABBBCCA$" is a minimum circular subsequence with a cardinality of three. When several cell IDs appear in a minimum circular subsequence of low cardinality, chances are that they are co-located or close to each other, and therefore may be put in the same cluster.

4.1   Naive Clustering Algorithm

A naive clustering algorithm is presented in Algorithm 1. The inputs are the cell ID sequence $\{(C_i, T_i)\}$ and a minimum circular subsequence cardinality threshold S. The output is a set of cell ID clusters $\{CL_i\}$[1]. w is a sliding window of cell IDs, CL holds the final output, CL' is an intermediate variable, and z holds all the cell IDs that have been clustered so far.

The **for** loop in lines 2 – 11 scans in time-ascending order the input sequence and identify minimum circular subsequences which, after duplicate elements are removed, are stored temporarily in CL'. We adopt notations for list indexing similar to the Python language, where positive indices are from left to right and negative indices are from right to left. The **while** loop in lines 12 – 20 then iterates through CL' to merge together clusters that share common cell IDs. Finally the **for** loop in lines 21 – 25 generates solo-clusters for cell IDs that do not belong to any existing clusters, i.e. creating a new cluster for each such cell ID.

---

[1] Cell ID clusters $\{CL_i\}$ are technically different from places $\{P_j\}$. Places are selected cell ID clusters that meet certain criteria described in Section 3.3.

Algorithm 1. `NaiveCluster()`: Cluster cell IDs based on cardinality threshold.

Require: $\{(C_i, T_i) \mid i = 0, 1, \ldots, N - 1; T_j < T_{j+1} \text{ for } j = 0, 1, \ldots, N - 2\}, S \geq 2$
Ensure: $\{CL_i \mid i = 0, 1, \ldots, N' - 1; CL_i \neq \Phi; \text{ all } CL_i \text{ form a partition of set } \{C_i\}\}$
 1: $w \leftarrow [], CL \leftarrow \Phi, CL' \leftarrow \Phi, z \leftarrow \Phi$
 2: **for** $i = 0$ to $N - 1$ **do**
 3:   **if** C_i not in w **then**
 4:     $w.append(C_i)$
 5:   **else if** $w[-1] = C_i$ **then**
 6:     **continue**
 7:   **else**
 8:     $j \leftarrow w.index(C_i), CL'.add(set(w[j:])), w \leftarrow w[0:j], w.append(C_i)$
 9:   **end if**
10:   $w \leftarrow w[-S:]$
11: **end for**
12: **while** CL' not empty **do**
13:   $a \leftarrow$ randomly selected element from $CL', CL'.remove(a), z \leftarrow z \cup a$
14:   **for all** b in CL' **do**
15:     **if** $a \cap b \neq \Phi$ **then**
16:       $a \leftarrow a \cup b, CL'.remove(b)$
17:     **end if**
18:   **end for**
19:   $CL.add(a)$
20: **end while**
21: **for all** C_i **do**
22:   **if** $C_i \notin z$ **then**
23:     $CL.add(set([C_i]))$
24:   **end if**
25: **end for**
26: **return** CL

Table 1. An example of the naive clustering algorithm

Cell ID sequence	w	CL'	CL
$\check{}AAABBCCCBDCD$	$[]$	$\{\}$	$\{\}$
$\check{A}AABBCCCBDCD$	$[A]$	$\{\}$	$\{\}$
$AAA\check{B}BCCCBDCD$	$[AB]$	$\{\}$	$\{\}$
$AAABB\check{C}CCBDCD$	$[BC]$	$\{\}$	$\{\}$
$AAABBCCC\check{B}DCD$	$[B]$	$\{\{B, C\}\}$	$\{\}$
$AAABBCCCB\check{D}CD$	$[BD]$	$\{\{B, C\}\}$	$\{\}$
$AAABBCCCBD\check{C}D$	$[DC]$	$\{\{B, C\}\}$	$\{\}$
$AAABBCCCBDC\check{D}$	$[D]$	$\{\{B, C\}, \{C, D\}\}$	$\{\}$
$AAABBCCCBDCD\check{}$	$[]$	$\{\}$	$\{\{B, C, D\}, \{A\}\}$

4.2 An Example

Table 1 gives an example on Algorithm 1. Input cell IDs are lined up in time-ascending order[2]. The cardinality threshold $S = 2$. The check mark (ˇ) indicates the current cell ID being looked at[3]. w, CL' and CL are shown as the algorithm progresses. Eventually B, C and D are clustered together; A forms a solo-cluster.

5 Algorithmic Improvements

5.1 Avoiding Over-Clustering

One problem with Algorithm 1 is the lack of resistance to the "ripple effect" introduced in the cluster-merging code in lines 15–17. To better understand it we refer to an example in Figure 1, in which a number of cell locations from the Open Cell ID project are mapped on Google Earth. Many cell towers are deployed along freeways and major roads; under circumstances such as stop-and-go traffic, it is not rare to see a mobile phone bouncing between cell IDs. Let us assume hypothetically that along a road there are a series of cells A, B, C, D and E. Since the input cell ID sequence to Algorithm 1 is from a long period of time, chances of seeing $\{A, B\}$, $\{B, C\}$, $\{C, D\}$ and $\{D, E\}$ in CL' may become non-negligible, eventually generating $\{A, B, C, D, E\}$ in CL. Consequently all five cell IDs are seen to represent the same place, which is a mistake.

Algorithm 2. QualifiedSet(): Compute cell IDs that can be clustered around.

Require: $\{(C_i, T_i) \mid i = 0, 1, \ldots, N-1; T_j < T_{j+1} \text{ for } j = 0, 1, \ldots, N-2\}, Q$
Ensure: $\{C_i'\}$ where C_i' appears at least Q times for at least one day.
  # Implementation omitted

The improvement is to allow clustering only around "qualified" cell IDs. For this purpose we introduce a function QualifiedSet() in Algorithm 2. It counts how many times each cell ID appears, on a daily basis, and returns those of which the number is above a given threshold Q for at least one day in the data set. Cell IDs returned from QualifiedSet() have demonstrated sufficient exposure to be considered for clustering. Algorithm 1 is then amended as in Algorithm 3.

5.2 Handling Missing Data

A second problem with the naive algorithm is its way of handling missing data. An implicit assumption in Algorithm 1 is that cell IDs are observed periodically, i.e. $T_{i+1} - T_i \approx \tau$ where $1/\tau$ is the sampling frequency. However there is no

---

[2] Throughout the paper we often use this abbreviated form of C_i in lieu of (C_i, T_i), and assume cell IDs to be in time-ascending order from left to right.

[3] Some intermediate steps are omitted for conciseness.

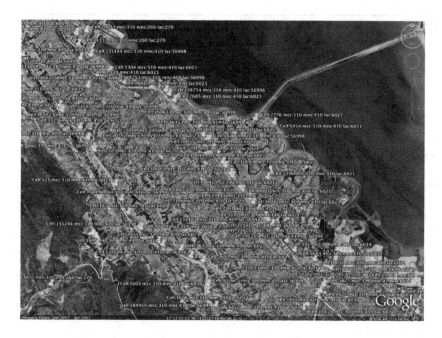

Fig. 1. Cell locations from the Open Cell ID project

Algorithm 3. Amendment to Algorithm 1.

 # The following line replaces the **Require** *line.*
Require: $\{(C_i, T_i) \mid i = 0, 1, \ldots, N - 1; T_j < T_{j+1}$ for $j = 0, 1, \ldots, N - 2\}, S \geq 2, Q$

 # The following line replaces line 1.
$w \leftarrow [], CL \leftarrow \Phi, CL' \leftarrow \Phi, z \leftarrow \Phi, QC \leftarrow QualifiedSet(\{(C_i, T_i)\}, Q)$

 # The following lines replace line 8.
$j \leftarrow w.index(C_i)$
if $set(w[j:]) \cap QC \neq \Phi$ **then**
$\quad CL'.add(set(w[j:]))$
end if
$w \leftarrow w[0:j], w.append(C_i)$

---

guarantee that a valid cell ID can always be observed steadily at this pace, e.g. the mobile phone may temporarily lose the signal or be turned off. Algorithm 1 must be further amended to accommodate such exceptions.

$$\ldots A????BA \ldots \tag{1a}$$

$$\ldots A????B????A \ldots \tag{1b}$$

Algorithm 4. Amendment to Algorithm 1, containing changes in Algorithm 3.

 # The following line replaces the **Require** *line.*
Require: $\{(C_i, T_i) \mid i = 0, 1, \ldots, N - 1; T_j < T_{j+1} \text{ for } j = 0, 1, \ldots, N - 2\}, S \geq 2, Q, \tau$

 # The following line replaces line 1.
$w \leftarrow [], w' \leftarrow [], CL \leftarrow \Phi, CL' \leftarrow \Phi, z \leftarrow \Phi, QC \leftarrow QualifiedSet(\{(C_i, T_i)\}, Q)$

 # The following lines replace lines 3–9.
if C_i not in w **then**
    $w.append(C_i), w'.append(T_i)$
else if $w[-1] = C_i$ **then**
    $w'[-1] \leftarrow T_i$
    continue
else
    $c \leftarrow 0$
    for all (w'_j, w'_{j+1}) in w' **do**
        if $w'_{j+1} - w'_j > 2\tau$ **then**
            $c \leftarrow c + 1$
        end if
    end for
    if $c > 1$ **then**
        $w \leftarrow [C_i], w' \leftarrow [T_i]$
    else
        $j \leftarrow w.index(C_i)$
        if $set(w[j :]) \cap QC \neq \Phi$ **then**
            $CL'.add(set(w[j :]))$
        end if
        $w \leftarrow w[0 : j], w.append(C_i), w' \leftarrow w'[0 : j], w'.append(T_i)$
    end if
end if

With extreme conservativeness we would tolerate no missing data, and any such event would cause the sliding window w to be reset; with extreme aggressiveness we would ignore the fact of missing data, and w would remain unchanged. To explain that neither case is desirable we refer to the examples in (1), in which a question mark is inserted into the cell ID sequence where an observation should have been recorded but is missing. (1a) has one chunk of missing data while (1b) has two. Since in (1a) there is evidence that B and A are adjacent, we argue that it is reasonable to cluster them together, if other conditions are met, even there is missing data between the first appearance of A and B. In contrast, we cannot rule out the possibility that B and A are distant from each other in (1b) because the observations are missing both *before* and *after* B, therefore we should reset w upon seeing the second missing data chunk.

We further amend Algorithm 1 to handle missing data as in Algorithm 4, which contains cumulatively all changes made in Algorithm 3. In a nutshell, a new list w' is added to track the timestamps in parallel to cell IDs. Before adding a cluster candidate to CL', the algorithm checks if there are two or more chunks of missing data in it, and drops the candidate if so.

6   Evaluation

6.1   Data Collection

We use the Nokia Simple Context [15] tool for data collection. The author has installed the Nokia Simple Context client on his primary mobile phone, a Nokia E71, and collected data for eight consecutive weeks. The phone was turned on most of the time during the day and turned off normally at night. In addition to the GSM cell information, Nokia Simple Context also collects a variety of data on GPS, Bluetooth, Wireless LANs, etc.; however we only focus on GSM data in this paper and leave study of multimodal localization for future work.

At the sampling rate of one per minute, a total of 54,842 GSM cell data points were recorded during the period, giving a coverage of about 914 hours[4]. Each data point consists of a Unix timestamp and time zone information, an ID that uniquely identifies the cell globally, the received signal strength and the number of signal bars. Figure 2 shows such a GSM cell data point in the JSON format. The author did not keep a diary on the ground truth, but most of it can be recovered easily from the author's calendar, email history, working notes and other sources.

{"tz": 28800, "user_id": xxxx, "type": "gsm", "source": "35292502072xxxx",
  "db_key": 4390xxxx, "time": 123991xxxx.0, "ver": "0.10",
  "data": {"cell_id": 16xxxx, "network_code": 410, "area_code": 56998,
          "signal_dbm": 95, "country_code": 310, "signal": 7}}

Fig. 2. A sample GSM cell data point (certain fields masked)

6.2   Distribution of Cell IDs

There are a total of 596 unique cell IDs observed in the data set. We plot the distribution of their appearances as the purple line in Figure 3. Cell IDs are sorted on the number of appearances and re-numbered before being plotted. Not to our surprise, it approximately follows the power law distribution [5]. The top two cell IDs apparently correspond to "Home" and "Work", respectively, given the author's lifestyle.

We then cluster the cell IDs according to Algorithms 1 – 4. We choose to set parameters $S = 2$, $Q = 10$ and $\tau = 5$ min. From experiments we find out that $S = 2$ is effective while larger values tend to over-cluster. $Q = 10^5$ has

---

[4] GSM cell information may occasionally be unavailable in Nokia Simple Context, in which case no data are recorded.

[5] Given the sampling rate of one per minute, it corresponds to 10 minutes.

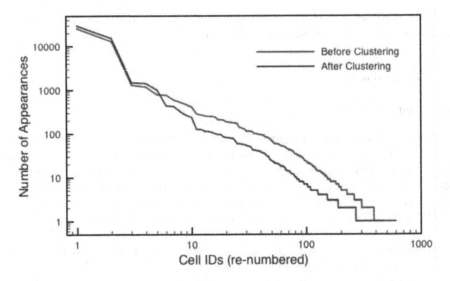

Fig. 3. Distributions of Cell ID appearances – before vs. after clustering

been shown [1][7] to be an empirically good value in location clustering. $\tau = 5$ min is chosen based on our observation that most temporary losses of cell ID information in the data set are less than 5 minutes.

Distribution of cell ID appearances after clustering is also plotted in Figure 3 as the blue line. The number of unique clusters is reduced to 472, but the line still resembles the power law distribution. All results hereafter in the paper are based on cell ID clusters rather than observed raw cell IDs.

Figure 4 plots the cell ID cluster sizes; clusters are first sorted in size-descending order before being plotted. The largest cluster consists of 21 cell IDs. There are a total of 30 clusters that consist of two or more cell IDs each; the rest are solo-clusters. We will revisit the cluster sizes shortly.

6.3 Significant Places – Overall

As we have pointed out, not every cell ID cluster represents a significant place. A cell ID cluster may potentially represent a significant place if it has been observed at least Q times on one or more days. We plot all such clusters in Figure 5, in which a pair of coordinates (x, y) means the cluster has been observed at least Q times on y days, while the daily average observation time is x. Generally speaking, a larger y suggests a stronger recurring pattern, while a larger x indicates more importance on certain days.

While sophisticated machine learning could be applied to Figure 5, it is simple to just identify points with a large x or y, or both. They correspond to the three types of significant places discussed in Section 3.3. We pick eight clusters, labeled A to H, and check against the author's calendar history to verify that they are indeed significant places. The result is shown in Table 2. The table is not meant

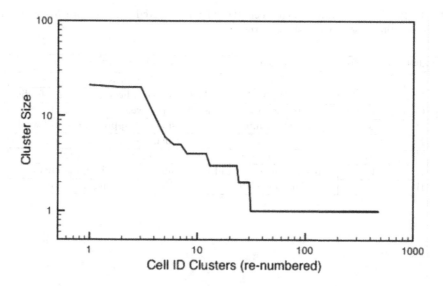

Fig. 4. Cell ID cluster sizes

to be exhaustive – there may exist other significant places unlabeled in Figure 5 and need more sophisticated algorithms to discover. In other words, the precision is high but the recall may be low. We leave a complete solution of the issue for future work.

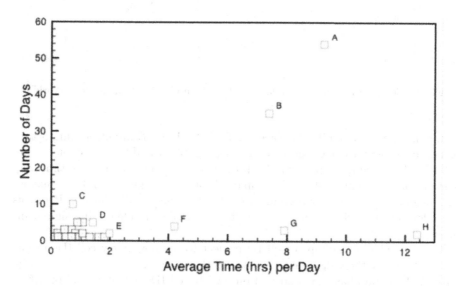

Fig. 5. Cell ID clusters that potentially represent significant places to the user

Table 2. Ground truth of eight cell ID clusters selected from Figure 5 and their corresponding places and visit patterns

Label	Cluster Size	Place	Normal Visit Pattern
A	21	Home	Daily unless out of town
B	20	Work	Daily on weekdays unless out of office
C	5	Grocery Store	Once per week unless out of town
D	3	Gymnasium	Occasionally, approx. once per week
E	4	UCLA	Business trip, long distance
F	10	Stanford Univ.	Business trip, local
G	6	Hotel in Southern Calif.	Business trip, long distance
H	20	Monterey, Calif.	Vacation, long distance

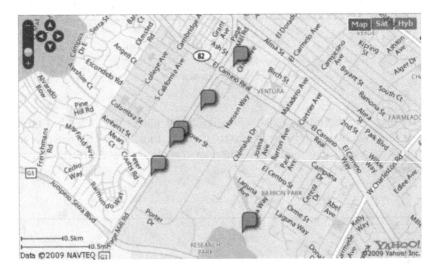

Fig. 6. Cell locations where available in the Open Cell ID database for "Work"

Among the eight significant places in Table 2, A and B are utterly important places to the user, while C and D are less frequently visited but still demonstrate a fairly clear recurring pattern. E through H, in contrast, are one-time trip destinations. It is interesting to compare Table 2 to Figure 4 and see that all eight significant places correspond to relatively-large cell ID clusters. It proves the relevance and efficacy of our clustering algorithms in identifying significant places.

We now refer to the Open Cell ID project to see how many cell IDs in our data set are in their database and where they are. As of June 2, 2009, the Open Cell ID database has information on 131 cell IDs out of a total of 596, a mere coverage of 22%. It shows that crowd-sourced cell-ID-to-physical-location mapping databases are still in an early stage. Among the eight significant places

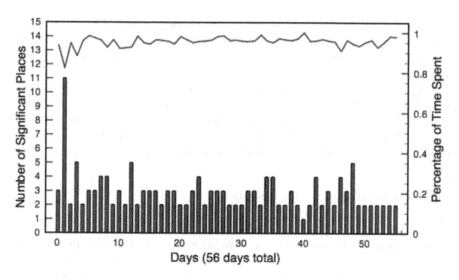

Fig. 7. Number of significant places – daily statistics

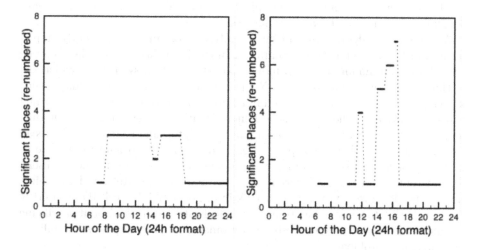

Fig. 8. A typical weekday (left)/weekend (right) with places and transit

in Table 2, "Work" has the best coverage with six cell IDs, plotted on the map in Figure 6. These cells span an area of approximately one mile in diameter, in line with our expectation on accuracy.

6.4 Significant Places – Daily Statistics

A basic unit of time in human life is day. In Figure 7 we show 1) how many significant places the author has for each day (brown bars), and 2) how much time the author spends in these places combined (purple line). The figure shows that for the vast majority of days, there are only 2 – 5 significant places. Day 2 is an exception on which the author was out of town and visited several venues. It was indeed a special day.

Finally Figure 8 shows how a typical day during the week (left) or weekend (right) looks like. Thick solid lines represent places; dotted lines mean "in transit". Both days share one and only one common place – "Home". The weekday is simply "Home → Work → Lunch → Work → Home"; the weekend is more diversified with short visits to several shopping, dining and entertainment places.

7   Discussion and Conclusion

In this paper we have proposed a mass market solution on mobile phones to discover a user's significant places. Our solution is solely based on periodically sampled, timestamped cell IDs – available on virtually all mobile phones today. We have evaluated the solution with data collected by the author over eight weeks. Results show that we are able to identify all three types of significant places of our interest: utterly important places such as "Home" and "Work", less frequently visited but recurring places such as "Grocery Store" and "Gymnasium", and one-time travel destinations.

All clustering algorithms and data analysis programs are currently implemented in Python on a laptop computer. In the immediate future they may be ported with minimum efforts to mobile phone operating systems in which the Python language is supported, i.e. Nokia S60, and also in other languages such as Symbian C++. We are interested in investigating advanced machine learning and data mining methods to discover more location patterns on different time scales, e.g. morning vs. evening, or summer vs. winter, and to improve the recall rate of significant places. We plan to expand evaluation to multiple people as well, once such data sets become available, and to compare results across users. Transit routes between places are also a natural next step. Multimodal localization is another direction we plan to look into, i.e. combining cell ID with GPS, Wi-Fi and Bluetooth in a way to improve localization results without sacrificing the merits. Finally we are interested in automated labeling of places – a difficult but intriguing problem.

References

1. Ashbrook, D., Starner, T.: Using GPS to Learn Significant Locations and Predict Movement Across Multiple Users. Personal and Ubiquitous Computing 7(5), 275–286 (2003)
2. Carlotto, A., Parodi, M., Valla, M.: Proximity Classification for Mobile Devices Using Wi-Fi Environment Similarity. In: MELT 2008, San Francisco, CA, pp. 43–48 (2008)
3. Djuknic, G.M., Richton, R.E.: Geolocation and Assisted GPS. ACM Computer 34(2), 123–125 (2001)

4. Dufkova, K., Ficek, M., Kencl, L., Novak, J., Kouba, J., Gregor, I., Danihelka, J.: Active GSM Cell-ID Tracking: Where Did You Disappear? In: MELT 2008, San Francisco, CA, pp. 7–12 (2008)
5. Faloutsos, M., Faloutsos, P., Faloutsos, C.: On Power-law Relationships of the Internet Topology. In: ACM SIGCOMM 1999, Cambridge, MA, pp. 251–262 (1999)
6. Hightower, J., Consolvo, S., LaMarca, A., Smith, I., Hughes, J.: Learning and Recognizing the Places We Go. In: Beigl, M., Intille, S.S., Rekimoto, J., Tokuda, H. (eds.) UbiComp 2005. LNCS, vol. 3660, pp. 159–176. Springer, Heidelberg (2005)
7. Kang, J.H., Welbourne, W., Stewart, B., Borriello, G.: Extracting Places from Traces of Locations. ACM Mobile Computing and Communications Review 9(3), 58–68 (2005)
8. Krumm, J., Horvitz, E.: LOCADIO: Inferring Motion and Location from Wi-Fi Signal Strengths. In: MobiQuitous 2004, Boston, MA, pp. 4–14 (2004)
9. Lassonen, K., Raento, M., Toivonen, H.: Adaptive On-Device Location Recognition. In: Ferscha, A., Mattern, F. (eds.) PERVASIVE 2004. LNCS, vol. 3001, pp. 287–304. Springer, Heidelberg (2004)
10. LaMarca, A., Chawathe, Y., Consolvo, S., Hightower, J., Smith, I., Scott, S., Sohn, T., Howard, J., Hughes, J., Potter, F., Tabert, J., Powledge, P., Borriello, G., Schilit, B.: Place Lab: Device Positioning Using Radio Beacons in the Wild. In: Gellersen, H.-W., Want, R., Schmidt, A. (eds.) PERVASIVE 2005. LNCS, vol. 3468, pp. 116–133. Springer, Heidelberg (2005)
11. Liao, L., Fox, D., Kautz, H.: Extracting Places and Activities from GPS Traces Using Hierarchical Conditional Random Fields. The International Journal of Robotics Research 26(1), 119–134 (2007)
12. http://www.google.com/latitude
13. http://www.navizon.com/
14. http://www.opencellid.org/
15. http://www.simplecontext.com/
16. http://www.skyhookwireless.com/

Adaptive Motion Model for a Smart Phone Based Opportunistic Localization System

Maarten Weyn[1,2], Martin Klepal[3], and Widyawan[3]

[1] Artesis University College of Antwerp,
Department of Applied Engineering: Electronics-ICT, Belgium
maarten.weyn@artesis.be
[2] University of Antwerp,
Department of Mathematics and Computer Science, Belgium
[3] Cork Institute of Technology, Ireland
{mklepal,widyawan}@cit.ie

Abstract. Localization systems will evolve towards autonomous system which will use any useful information provided by mobile devices taking the hardware specification and environmental limitations into account. This paper demonstrates the concept of opportunistic localization using a smart phone with the following sensor technologies: Wi-Fi, GSM, GPS and two embedded accelerometers. A particle filter based estimator with an adaptive motion model is used to seamlessly fuse the different sensory readings. Real experiments in multi-floor, indoor-outdoor environments were conducted to analyze the performance of the proposed system. The achieved results using various sensor combinations are presented.

1 Introduction

The future of localization systems most likely will evolve towards systems which can adapt and cope with any available information provided by mobile clients. However, one of the common disadvantages of many existing localization systems is the need for dedicated devices and proprietary infrastructure in the operation area of the indoor localization system.

The increasing proliferation of mobile devices, such as PDA's and smart phones, has fostered growing interest in location based applications which can take advantage of available information in the environment that can be extracted by the mobile device. This type of localization is called opportunistic localization. For example, most mobile devices can provide GSM related data like the connected cell tower identification and signal strength, whereas more advanced devices are equipped with Wi-Fi, GPS or a combination of the previous. More recent mobile devices also have inertial sensors build in (mostly accelerometers) which can be used as extra localization information using the Pedestrian Dead Reckoning (PDR) [1] principle, where the internal accelerations are used to estimate the displacement of the person.

The purpose of this paper is to present the Opportunistic Seamless Localization System (OSL), a smart phone based opportunistic localization system

R. Fuller and X.D. Koutsoukos (Eds.): MELT 2009, LNCS 5801, pp. 50–65, 2009.
© Springer-Verlag Berlin Heidelberg 2009

which combines sensor data available from mobile devices. An adaptive motion model for the particle filter is proposed to seamlessly fuse different technologies based on the availability of accelerometer and GPS data from the smart phone.

1.1 Related Works and Contribution

As shown in [2] a lot of techniques can be used for localization. GSM is a widely used technology for the localization of handheld devices. PlaceLab [3], probably the most know system which combines different technologies, combines GPS, Wi-Fi and GSM beacons. Systems using Wi-Fi alone like [4] have already proven themselves, but are very depended on the Wi-Fi access point placement and RF-fingerprint, since they only use Wi-Fi and do not integrate any other sensors.

In OSL, GSM is not used for cell ID localization as in PlaceLab, but the signal strength measurements are directly used in the measurement model of the particle filter, similar to our Wi-Fi implementation, to be compared to the RF-fingerprint database. Accelerometers are already used in other projects like [1], but all other research on PDR mostly uses dedicated accelerometers, where OSL uses the internal sensors of the smart phone. This also means that the algorithm should cope with the placement of the smart phone, which is for example in a trousers pocket in an unknown orientation, where most other research used accelerometers place on a special part of the body (foot, chest ...).

In this paper we want to focus on the novel implementation of the adaptive motion model, which will dynamically use the available opportunistic sensor data.

The remainder of the paper is organized as follows. In Section 2, the architecture of the OSL system is introduced. The fusion algorithm and the adaptive motion model will be proposed in the Section 3. Section 4 will describe the experiment and results. Finally, section 5 concludes the paper.

2 Architecture

The current version of OSL has an event-driven client-server based architecture, where the client consists of different sensor services dependent on the hardware capabilities. These sensors send events to a central client component which will handle all events and initiate the communication with the server. The server, besides the data fusion, also handles the preprocessing of this data for the fusion engine and the post-processing for the visualization and logging.

2.1 The Mobile Devices

In the current prototype, the client is an Openmoko Neo Freerunner PDA. The PDA is able to provide the following data:

- GPS data
- GSM Cell readings (serving and up to 8 neighboring cells)
- Wi-Fi RSSI measurements
- Step detection data and stride estimation based on 2 accelerometers

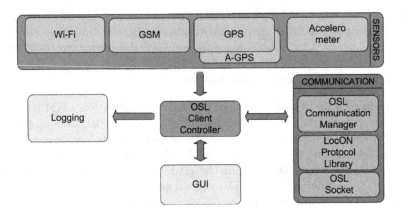

Fig. 1. The Client Architecture

The client can work in 2 different modes: an online mode and a logging mode. In the *online mode*, there is a connection between the client and the server. The client sends all required data to the server. Which data is required, is determined in collaboration with the server. In a later phase, the server tells the client which data to send in order to optimize performance and minimize the data throughput. In the *logging mode*, the client is not connected to the server and saves all data which will be send during the next connection to the server.

As shown in Figure 1 the client consists of different components: the sensors, the communication, the controller, the logging component and the graphical user interface (GUI), the first three are described below. The client is running QtExtended 4.4.3 on Linux and all components are written for this system, which makes its more portable to other platforms.

2.2   The Mobile Device Sensors

The number of sensors is dependent on the client's hardware specifications; in our system, the PDA has the following sensors GPS, Wi-Fi, GSM and 2 accelerometers:

GPS. The software daemon parses the data coming from the GPS [5] chip, and sends the useful data to the OSLClient Controller. Currently the following NMEA (National Marine Electronics Association) messages are used and parsed:

- $GPGSA: the GPS Dilution of Precision DOP (Horizontal HDOP, 3D PDOP and Vertical VDOP) and information about the active satellites. The DOP is an indicator of the geometric configuration of the active satellites, which influences the GPS accuracy.
- $GPGGA: the position, HDOP, number of satellites, the fix quality and the altitude.
- $GPZDA: the date and time, this is used to synchronize the PDA's time.

– $GPRMC: the required minimum specific GPS data, which is the position and speed, the speed is received in knots but is converted to km/h.

The combination of these messages gives the position, speed and altitude, together with quality information like the different dilution of precisions and the number of active satellites.

Assisted-GPS. The issue, which had to be overcome when using the GPS chip used in Neo Freerunner (u-blox ANTARIS 4) is that it has no internal memory to store known almanac data and the latest known position and time, which causes a long Time To First Fix (TTFF) for every new cold start. Currently the Assisted-GPS (AGPS) module is used every time to speed up the start up time.

The position used to get the AGPS data from the u-blox AGPS server can be a rough estimate from the last known location or the position calculated from Wi-Fi or GSM, similar to [6].

Wi-Fi. Localization using Wi-Fi, is probably the most commonly used technology for indoor localization as it benefits of the omnipresence of Wi-Fi networks. Nearest access point (AP), multi-lateration[7] and fingerprinting[8] can be used to determine a position. Since the idea of OSL is to create sensors which can be re-used as much as possible, multi-lateration using timing is quite difficult to achieve since it is not feasible to get precise timing information using standard devices in a generic way. This makes time dependent localization techniques like TDOA and TOA not possible to use for generic PDA's. In OSL fingerprinting using signal strength is implemented, this requires a fingerprint database for Wi-Fi, which in the latest version should be self adopting using a self-calibration algorithm.

GSM. Every mobile phone has the identification and signal strength of the cell tower to which it is connected (serving cell). A broadcast message of this tower informs all connected cell phones about eight neighboring cell towers which can be used in case a handover is needed. The measured GSM/UMTS signals strengths from this serving and up to eight neighboring cells are treated similarly to Wi-Fi signal strength discussed above. Cell-ID localization is used in different case studies[9,10] and applications already. In Cell-ID localization the mobile phone is located in the area covered by the connected cell tower. The implementation of Cell-ID localization adds straightforward coarse-grained indoor and outdoor location information to OSL, but the location of all cell towers is needed. However, the main benefit of GSM is in urban (near-) indoor environments where the pattern, created from the GSM field strengths of all cell towers, can provide localization where GPS fails due to the blockage of the satellite signals (shadowing), similar to [11]. Thanks to the influence of for example walls on the GSM field strength the fingerprint build with the field strengths of the different cell towers differs significantly in different locations of a building. These differences are used in localization using fingerprinting or pattern matching like in [4]. The difference in OSL towards other research on indoor GSM localization like [11], is that a standard PDA does not allow getting system information of all surrounding cell towers. Since we try to built OSL as generic as possible we can only use the information which is standard available in most mobile phone. Therefore, OSL uses a standard GSM fingerprinting to cover areas with poor GPS and Wi-Fi coverage where the main power lies in the algorithm using the different field strength measurements.

Mobile Client Motion Detection. The accelerometers are used to process step-detection using the algorithm described in [12]; the number of steps will be send to the server if they can be calculated, together with an estimated distance. This sensor gives us not only possible extra knowledge about the traveled distance, which is used for PDR, but as well if the person, using the device, is moving or not. This information will be used in the motion model described below.

2.3   The OSL Client Controller

The controller is responsible to collect all events coming from the sensors and sent them to the communication module, which sends it to the server. This controller is adapted according to the devices hardware specification to allow all possible sensory data to be used. The controller also interacts with the GUI and the logging component.

2.4   The Communication

In a system which should support all kinds of devices, a standardized way to transfer the data from the client to the server is needed. That is why an open localization data stream binary protocol called LocON has been adopted. The LocON protocol was developed within the research conducted in the EC FP7 LocON project [13] for the transparent localization data streaming within localization system components. Security and encryption is also supported by this protocol. The communication component is responsible for queuing all data coming from the sensor, building the LocON compliant message and sending them to the sever.

3   Seamless Sensor Fusion

3.1   Sequential Non-linear Bayesian Filtering

A particle filter [14], a technique which implements the recursive Bayesian filtering using the sequential Monte Carlo method, is currently one of the most advanced techniques for sensor data fusion. A particle filter allows to model the physical characteristics of the movement of an object (motion model) and gives a weight (corresponding to the believe) to the particle using the noisy observations, the measurements (measurement model), of for example Wi-Fi, GSM and GPS. After each update step the particles are resampled according to their weight. The main advantage of particle filters against Kalman Filters, for example, is that it can deal with non-linear and non-Gaussian estimation problems. Additional information like map filtering, where environmental knowledge like the location of walls is incorporated in the motion model to remove impossible trajectories [4], is easily implemented using particle filters.

The particle filter directly estimates the posterior probability of the state $\mathbf{x}_t$, which is expressed with the following equation [15]:

$$p(\mathbf{x}_t|\mathbf{z}_t) \approx \sum_{i=1}^{N} w_t^i \delta(\mathbf{x}_t - \mathbf{x}_t^i) \tag{1}$$

where $\mathbf{x}_t^i$ is the i-th sampling point or particle of the posterior probability with $1 < i < N$ and w_t^i is the weight of the particle. N represent the number of particles in the particle set and $\mathbf{z}_t$ represents the measurement.

Algorithm 1 describes the generic algorithm of a particle filter. The input of the algorithm is the previous set of the particles $\mathcal{X}_{t-1}$, and the current measurement $\mathbf{z}_t$, whereas the output is the new particle set $\mathcal{X}_t$.

In our OSL system, the state $\mathbf{x}$ of a particle represents the position (x_t^i, y_t^i, z_t^i), its velocity v_t^i, bearing β_t^i and validity $valid_t^i$ (for example to identify if a particle has crossed a wall).

Algorithm 1. Particle_Filter ($\mathcal{X}_{t-1}, z_t$)

1: $\bar{\mathcal{X}}_t = \mathcal{X}_t = \emptyset$
2: **for** $i = 1$ to N **do**
3:     sample $\mathbf{x}_t^i \sim p(\mathbf{x}_t|\mathbf{x}_{t-1}^i)$
4:     assign particle weight $w_t^i = p(\mathbf{z}_t|\mathbf{x}_t^i)$
5: **end for**
6: calculate total weight $k = \sum_{i=1}^{N} w_t^i$
7: **for** $i = 1$ to N **do**
8:     normalize $w_t^i = k^{-1} w_t^i$
9:     $\bar{\mathcal{X}}_t = \bar{\mathcal{X}}_t + \{\mathbf{x}_t^i, w_t^i\}$
10: **end for**
11: $\mathcal{X}_t = $ **Resample** ($\bar{\mathcal{X}}_t$)
12: **return** $\mathcal{X}_t$

The algorithm will process every particle $\mathbf{x}_{t-1}^i$ from the input particle set $\mathcal{X}_{t-1}$ as follows:

1. Line 3 shows the prediction stage of the filter. The particle $\mathbf{x}_t^i$ is sampled from the transition distribution $p(\mathbf{x}_t|\mathbf{x}_{t-1})$. The set of particles resulting from this step has a distribution according to (denoted by $\sim$) the prior probability $p(\mathbf{x}_t|\mathbf{x}_{t-1})$. This distribution is represented by the motion model.
2. Line 4 describes the incorporation of the measurement $\mathbf{z}_t$ into the particle. It calculates for each particle $\mathbf{x}_t^i$ the *importance factor* or *weight* w_t^i. The weight is the probability of the received measurement $\mathbf{z}_t$ for particle $\mathbf{x}_t^i$ or $p(\mathbf{z}_t|\mathbf{x}_t)$. This is represented by the measurement model.

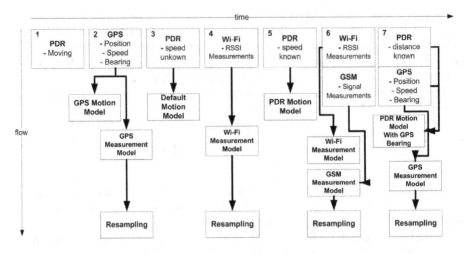

Fig. 2. An example of the fusion flow

3. Line 7 until 10 are the steps to normalize the weight of the particles. The result is the set of particles $\bar{\mathcal{X}}_t$, which is an approximation of posterior distribution $p(\mathbf{x}_t|\mathbf{z}_t)$.

4. Line 11 describes the step which is known as *resampling* or *importance resampling*. After the resampling step, the particle set which is previously distributed equivalent to the prior distribution $p(\mathbf{x}_t|\mathbf{x}_{t-1}, \mathbf{z}_{t-1})$ will be changed to the particle set $\mathcal{X}_t$ which is distributed in proportion to $p(\mathbf{x}_t|\mathbf{x}_{t-1}, \mathbf{z}_t)$.

3.2 Fusion Flow

Figure 2 shows an example on how data is handled when it arrives at the fusion engine. From left to right the data arriving at different time stamps are shown. The way how this data is handled, according to the different steps of a particle filter, is described below:

1. Knowledge about moving or standing still is saved to be used during later decisions.

2. Data from GPS can be used for the motion model (since we have an idea of the approximate speed and bearing), and afterwards in the GPS measurement model. After the measurement model, the particles are resampled.

3. If the PDR tells the engine that there was movement for a certain time, but is unable to estimate the traveled distance, the standard motion model will be used.

4. The Wi-Fi data can be used directly in the Wi-Fi measurement model, which is followed by the resampling.

5. If we get the estimated distance from the PDR, this is used in the PDR motion model.

6. If Wi-Fi and GPS data are arriving at the same time interval, the measurement models will be used directly after each other, which will multiply the two likelihood functions. After processing both, the resampling will select the samples with the best probability.
7. If PDR and GSM data is arriving together, the PDR motion model will use the GPS bearing as well. After which the GPS measurement model and the resampling is done.

3.3   Motion Model

The motion model incorporating the input from PDR, GPS and the map filtering is shown in Algorithm 2, where $\mathbf{x}_{t-1}^i$ is the previous state of the particle, $\mathbf{u}_t$ is the control or action (PDR, GPS or unknown) and Δ_t is the time difference between the two update steps.

Algorithm 2 describes the dynamic motion model, which will use different equations to determine the next position depending on the control data. The motion models currently integrated in the adaptive motion model are a default motion model, a GPS motion model and a PDR motion model.

- The algorithm will try maximum *maxRetries* times (in our implementation 5) to calculate a new possible location for the particle, taking into account all the control data and the environment (walls).
- First the new speed will be calculated. If we know that the object is moving (we assume it is always moving if we have no information coming from the PDR) and there is speed information coming from PDR, a speed will be randomly chosen from the Gaussian distribution specified by mean v_{pdr} and standard deviation σ_v, $\frac{1}{20}$ is related to the accuracy of the PDR algorithm [12]. If there is no information from PDR, but there is information coming from GPS, we will similarly get a speed from the GPS sensor.
  If we have no information at all about the real speed, a new speed will be chosen dependent on the previous speed of the particle. Thanks to the resampling depending on the measurement model only particles with a realistic position, hence a corresponding speed will be chosen.
  If we have information coming from the PDR that the object is not moving, the speed will be chosen to be $0m/s$.
  The resulting speed is limited to v_{max}, in our case $10m/s$ for people.
- In the current OSL implementation we can only get direction information from the GPS sensor. If we have this information we will randomly choose the bearing from the Gaussian distribution specified by a mean being the average between the bearing coming from GPS and the previous bearing and the standard deviation dependent on the speed. The reason is that the deviation by the object will be very small if it is moving fast, and can be bigger if the object is moving very slow.
  If no information from GPS is available the mean of the Gaussian distribution will only depend on the previous bearing of the particle.
  The angle depicting the bearing is limited between $-pi$ and pi.

Algorithm 2. Motion_Model (x_{t-1}^i, u_t, Δ_t)

1: **for** $i = 1$ to $maxRetries$ **do**
2:   **Speed**
3:   **if** pdr_{moving} **then**
4:     **if** $pdr_{speedknown}$ **then**
5:       $v_t = N(v_{pdr}, \sigma_v)$, where $\sigma_v = \frac{1}{20}.\sqrt{\Delta_t}$
6:     **else**
7:       **if** $gps_{speedknown}$ **then**
8:         $v_t = N(v_{gps}, \sigma_v)$, where $\sigma_v = \frac{1}{5}.\sqrt{\Delta_t}$
9:       **else**
10:        $v_t = N(v_{t-1}, \sigma_v)$, where $\sigma_v = 1.\sqrt{\Delta_t}$
11:      **end if**
12:    **end if**
13: **else**
14:    $v_t = 0$
15: **end if**
16: with $\{ \begin{array}{l} v_t = |v_t|, \text{ if } v_t < 0 \\ v_t = 2.v_{max} - v_t, \text{ if } v_t > v_{max} \end{array}$
17: **Bearing**
18: **if** $gps_{bearingknown}$ **then**
19:    $\beta_t = N(\sqrt{\beta_{gps}.\beta_{t-1}}, 0.5\pi - \arctan(\frac{\sqrt{v_{gps}}}{2}))$
20: **else**
21:    $\beta_t = N(\beta_{t-1}, 0.5\pi - \arctan(\frac{\sqrt{v_{gps}}}{2}))$
22: **end if**
23: with $\{ \begin{array}{l} \beta_t = \beta + 2\pi, \text{ if } \beta_t < -\pi \\ \beta_t = \beta_t - 2\pi, \text{ if } \beta_t > \pi \end{array}$
24: **Position**
25: $\sigma_p = 0.5m$
26: $X_t = [\begin{smallmatrix} x_t \\ y_t \end{smallmatrix}] = [\begin{smallmatrix} x_{t-1} + v_t.\cos(\beta_t).\Delta_t + n \\ y_{t-1} + v_t.\sin(\beta_t).\Delta_t + n \end{smallmatrix}]$, where $n = N(0, \sigma_p)$
27: **Floor Change**
28: **if** ParticleCrossedFloorChangeArea **then**
29:    Change z_t
30: **end if**
31: **Map Filtering**
32: $valid$ =checkIfPartcleIsCrossingAWall
33: **if** valid **then**
34:    **return** x_t^i
35: **end if**
36: **end for**
37: **return** x_t^i

- Finally the new position (x_t and y_t coordinate) is calculated using the previous position and the speed and bearing discussed above. Gaussian noise n is added.
- If the environment specifies transition areas between floors, we can calculate if the new particle position is located in an area where a transition to another floor is possible (for example stairs or elevators).

– With the new position known, we now check if the particle made a valid move from the former position to the new position (if it didn't cross a wall). If it was a valid transition, the new particle position is returned, otherwise a new position is being calculated. If after *maxRetries* still no valid position can be found, the particle's validity will be set to zero so the weight of the particle can be set to zero in the measurement model.

The process of removing incorrect particles is called map filtering, since the map is known to visualize the object afterwards, wall information can in most cases be extracted.

3.4   Measurement Model

The Wi-Fi and GSM measurement model algorithm returns for each particle a weight corresponding to the likelihood function, which is build using the fingerprints. The weight represents the correspondence between the measurement and the fingerprint at the location of the particle, similar to [16]. The GPS measurement model returns a weight corresponding to the difference between the particle position and the GPS position taking the GPS internal error and the delusion of precision into account.

4   Experiment and Result

Figure 3 shows a part of the CIT buildings where all test runs where done. Every floor is 3.65m high, the corridors are 2.45m width and the part of the floors which is used is 44m wide and 110m long. A 'quick-and-dirty' fingerprint was made for Wi-Fi and GSM, in the corridor and the rooms which where accessible. The fingerprinting was done 2 months before the actual test measurement which is described in this paper, to include the influence of signal changes due to environmental changes. The Wi-Fi access points which were used, were placed there for a data network, not for localization, which has the effect that in most places only 2, sometimes 3 access points are visible, with a non-ideal geometric configuration. The placement of the access points are shown in Figure 3 with a triangle. If the localization accuracy of Wi-Fi should be improved, extra access points could be added, taking into account the geometric diversity of the access point locations. The locations of the GSM cell towers are unknown since we do not want to make the system dependent on this information.

The fingerprinting of Wi-Fi and GSM was done while holding the Neo PDA, to be able to insert reference points by touching on the screen the current place on a detailed map. This fingerprinting could be improved by taking more measurements to collect more data to incorporate for example all different antenna orientations of the PDA. But the main goal is to test the opportunistic localization by using 'non ideal data'. The measurement was done with another Neo PDA having the PDA in the trousers pocket, which influences the signal strength heavily, again to stress the 'non ideal', opportunistic concept.

Different test runs with other trajectories where made which all had comparable results. The trajectory discussed in this paper can be seen in Figure 3.

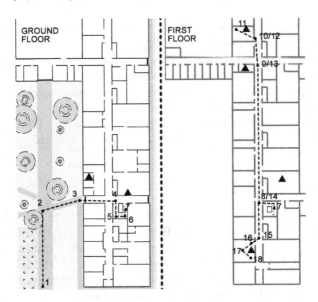

Fig. 3. The part of 2 floors of the CIT building with the track of the test run, described in this paper

This trajectory is chosen because it demonstrates the outdoor-indoor seamless transitions and the transition between floors. The test person came from outdoor, walked into the corridor of the ground floor, took the stairs to the first floor, walked using the corridor to an office, turned there and walked using the same corridor to another office. The person was constantly in movement, which will make it harder for the fusion engine to estimate the location.

4.1   Results

Figure 4 and 6 and Table 3 give an overview of the mean estimation error, split up in the different areas (outdoor, first floor and second floor). All systems use map filtering. As expected is the fusion using all possible information (*Wi-Fi+GSM+GPS+PDR*) giving the best results. The improvement using PDR towards the fusion without PDR (*Wi-Fi+GSM+GPS*) is mostly visible in the indoor part, since in outdoor, the speed and bearing of GPS is used in the motion model.

Wi-Fi performs best inside, while GPS only works outside and in the first few meters while walking inside. Since the Neo PDA does not contain an ultra sensitive GPS receiver and the GPS signals are not able to penetrate the roof and walls enough to be received by the PDA, GPS cannot be used on any other indoor location in this test. Furthermore the PDA is located in the trousers pocket of the test person, which makes it even harder to receive any GPS signals. GSM localization using fingerprinting (pattern matching) will work best indoors, but is most powerful in combination with Wi-Fi since their likelihood functions will mostly complement each other.

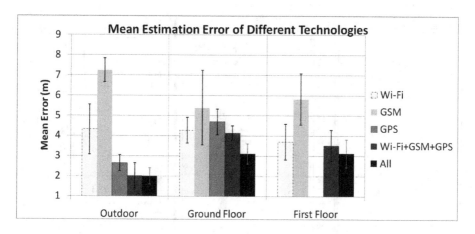

Fig. 4. Results of location estimation using different technologies

Table 1. Example GSM Measurement

Cell Tower ID	Field Strength
5-37231-789	-90 dB
24-25025-17	-100 dB
3-27642-799	-106 dB
3-2764-20	-100 dB

Table 2. Example Wi-Fi Measurement

Wi-Fi Access Point	Signal Strength
00:20:A6:62:87:1B	-61 dB
00:20:A6:63:1D:4F	-93 dB

Figure 5 shows two consecutive measurements, the left one a PDR+GSM measurement and the right one PDR+Wi-Fi. The cross depicts the real position, the black dot the estimated position. The dark grey cloud are the particles, the light grey dots represent the likelihood observation function (LOF) of the measurement (more dense = more likely). The ground floor is not shown in the Wi-Fi figure since the LOF is equal to zero for all places on the ground floor.

The particles distribution at the left side (PDR+GSM) is the result of the position of the particles in the previous particle set, the motion model defined by PDR (moving = true, speed = 1.7m / 1.47s) and the measurement model defined by the GSM measurement (shown in Table 1) and the resampling step.

The particle distribution at the right side is formed starting from the particle distribution after the step described above, the motion model defined by PDR (moving = true, speed = 1.6m / 1.37s) and the measurement model defined by the Wi-Fi measurement (shown in Table 2) and the resampling step.

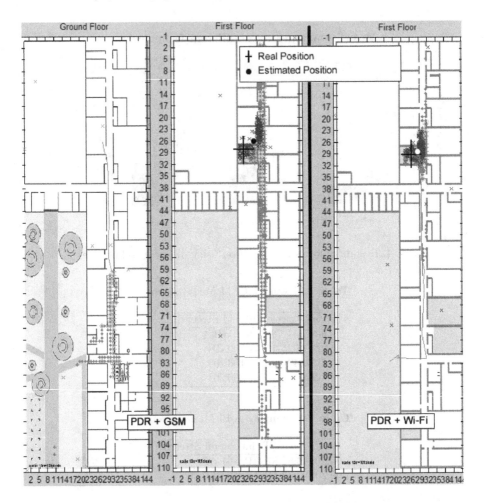

Fig. 5. Particle Filter of PDR+Wi-Fi and PDR+GSM measurement

It is also visible that GSM has more problems detecting on which floor the object is located then Wi-Fi. This is because the signal strengths of GSM on two floors directly under each other, does not differ very significantly if the floor construction is comparable with each other. Again the combination of both Wi-Fi and GSM gives an improvement.

The results also show that the addition of PDR give a slight improvement since the motion model guides the particles better towards the real position of the object, for example around the staircase where the particles can go from the ground to the first floor. The occurrence of estimating the wrong floor, only happens around the time where the object moves from one floor to the other. During normal movement of the object, the motion model does not allow switching floors if the particles are not in the neighborhood of a transition area (stairs or elevator).

Table 3. Results of location estimation using different technologies. (in meter)

Technology	Mean Error	Std. Dev.	Outdoor	Floor 0	Floor 1	Correct Floor
GPS	3.08	1.31	2.65	4.69	N/A	N/A
GSM	6.18	2.36	7.24	5.39	5.82	70%
Wi-Fi	3.90	1.88	4.32	4.27	3.70	88%
Wi-Fi+GSM+GPS	3.05	1.49	2.02	4.14	3.51	91%
Wi-Fi+GSM+GPS+PDR	2.73	1.28	2.01	3.10	3.12	93%

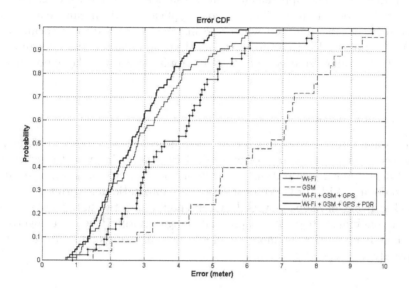

Fig. 6. CDF Error of location estimation using different technologies

The accuracy of the Wi-Fi and therefore also of the fusion using Wi-Fi to-gether with another technology, can be greatly improved by adding extra access points, since in most of the locations only 2 access points are visible. The accu-racy improves to about a meter if the person is standing still for a while, which gives the particles the opportunity to converge around the real location.

5   Conclusion

In this paper we have demonstrated the concept of opportunistic localization where the location estimation is done only with the available opportunistic data, depending on the environment and the client's hardware capabilities. A particle filter with an adaptive motion model is used to fuse the different sensor data. We have shown the resulting localization using a device which is only able to use a separate technology and the results if a device is able to collect more technologies so the fusion engine can combine them.

We have focused on the opportunistic concept, where the data measurement, fingerprinting and used hardware should represent realistic non-ideal data. Where the user is not interested in installing any extra hardware to be able to have a localization service, but where the service will adapt to the environmental opportunities. The tests in Cork Institute of Technology, Ireland, show localization using GSM with a mean error of 6.18m using standard known signals of up to 8 surrounding cell towers, for Wi-Fi around 3.90m where in most cases only 2 access points where visible. The fusion of GSM, Wi-Fi, GPS, PDR and map filtering, gives an improvement, which results in a mean error of 2.73m and a correct floor detection of 93%. In most application which can benefit from opportunistic localization and where the installation of dedicated localization hardware is not needed to improve the localization accuracy and a mean error around 3m will allow sufficed data to build application using the opportunistic data.

5.1   Further Work

OSL can be improved by adding sensor possibilities like Bluetooth which makes the detection of the proximity of other devices possible in order to implement mutual device aided localization. A the moment a fingerprint is still needed in order to do any Wi-Fi or GSM localization, an automatic fingerprint prediction method is being developed by CIT which should ease the initial installation. Another possibility is to create an automatic fingerprinting, where the fingerprint is continually updated when more useful information is coming to the server, this is currently developed by Artesis.

Acknowledgments

The research into OSL's core technology was conducted in the context of following projects: the EC FP7 LocON and the Enterprise Ireland's Proof of Concept.

References

1. Beauregard, S.: Omnidirectional pedestrian navigation for first responders. In: 4th Workshop on Positioning, Navigation and Communication, 2007. WPNC 2007, pp. 33–36 (2007)
2. Porretta, M., Nepa, P., Manara, G., Giannetti, F.: Location, location, location. IEEE Vehicular Technology Magazine 3(2), 20–29 (2008)
3. LaMarca, A., Chawathe, Y., Consolvo, S., Hightower, J., Smith, I., Scott, J., Sohn, T., Howard, J., Hughes, J., Potter, F., et al.: Place lab: Device positioning using radio beacons in the wild. In: Gellersen, H.-W., Want, R., Schmidt, A. (eds.) PERVASIVE 2005. LNCS, vol. 3468, pp. 116–133. Springer, Heidelberg (2005)
4. Widyawan, Klepal, M., Beauregard, S.: A novel backtracking particle filter for pattern matching indoor localization. In: MELT 2008: Proceedings of the first ACM international workshop on Mobile entity localization and tracking in GPS-less environments, pp. 79–84. ACM, New York (2008)

5. El-Rabbany, A.: Introduction to GPS: The Global Positioning System. Artech House (2002)
6. Weyn, M., Schrooyen, F.: A wifi-assisted-gps positioning concept. In: Proceeding of ECUMICT, Gent, Beglium (March 2008)
7. Wibowo, S., Klepal, M., Pesch, D.: Time of flight ranging using off-the-shelf wifi tag. In: Proceedings of PoCA 2009 (2009)
8. Widyawan, Klepal, M., Pesch, D.: A bayesian approach for rf-based indoor localisation. In: IEEE ISWCS, Trondheim, Norway (October 2007)
9. Wigren, T.: Adaptive enhanced cell-id fingerprinting localization by clustering of precise position measurements. IEEE Transactions on Vehicular Technology 56, 3199–3209 (2007)
10. Caffery, J., Stuber, G.: Overview of radiolocation in cdma cellular systems. IEEE Communications Magazine 36(4), 38–45 (1998)
11. Varshavsky, A., de Lara, E., Hightower, J., LaMarca, A., Otsason, V.: Gsm indoor localization. Pervasive and Mobile Computing 3(6), 698–720 (2007)
12. Bylemans, I., Weyn, M., Klepal, M.: Mobile phone-based displacement estimation for opportunistic localisation systems. In: Proceedings of UBICOMM9: International Conference on Mobile Ubiquitous Computing, Systems, Services and Technologie, Malta (October 2009)
13. Couronne, S., Hadaschik, N., Fassbinder, M., von der Grun, T., Klepal, M., Widyawan, W.M., Denis, T.: Locon a platform for an inter-working of embedded localisation and communication systems. In: Proceeding of 6th Annnual IEEE SECON 2009, Rome, Italy (June 2009)
14. Thrun, S., Burgard, W., Fox, D.: Probabilistic Robotics (Intelligent Robotics and Autonomous Agents). MIT Press, Cambridge (2006)
15. Ristic, B., Arulampalam, S.: Beyond the Kalman filter: Particle filters for tracking applications. Artech House (2004)
16. Evennou, F., Marx, F.: Advanced integration of WIFI and inertial navigation systems for indoor mobile positioning. EURASIP Journal on Applied Signal Processing 2006, 1–11 (2006)

Model-Free Probabilistic Localization of Wireless Sensor Network Nodes in Indoor Environments*

Ioannis C. Paschalidis[1,2], Keyong Li[1], and Dong Guo[1]

[1] Center for Information & Systems Engineering
[2] Dept. of Electrical & Computer Eng., and Division of Systems Eng.
Boston University, 15 St. Mary's St., Brookline, MA 02446
yannisp@bu.edu
http://ionia.bu.edu/

Abstract. We present a technique that makes up a practical probabilistic approach for locating wireless sensor network devices using the commonly available signal strength measurements (RSSI). From the RSSI measurements between transmitters and receivers situated on a set of landmarks, we construct appropriate probabilistic descriptors associated with a device's position in the contiguous space using a pdf interpolation technique. We then develop a localization system that relies on these descriptors and the measurements made by a set of clusterheads positioned at some of the landmarks. The localization problem is formulated as a composite hypothesis testing problem. We develop the requisite theory, characterize the probability of error, and address the problem of optimally placing clusterheads. Experimental results show that our system achieves an accuracy equivalent to 95% < 5 meters and 87% < 3 meters.

1 Introduction

A reliable indoor positioning service gives rise to a plethora of important applications ranging from asset tracking to disaster response. The GPS technology is hardly operational for indoor use. Many other ideas have been investigated, some of which may require special hardware/infrastructure. Our primary interest is in methods that allow us to add the positioning service to an existing wireless sensor network (WSNET), using only the basic measurements of the radio frequency (RF) communications in the WSNET — specifically, the measurements of the received signal strength indication (RSSI).

For a brief (hence, incomplete) review of the RF-based positioning literature, the systems proposed by [1, 2, 3, 4] compare mean RSSI measurements to a precomputed signal-strength map. These systems succeeded in demonstrating the feasibility of providing meaningful positioning services using WSNETs and injected enthusiasm into the field. However, their performances leave room for

---

* Research partially supported by the NSF under grants DMI-0330171, ECS-0426453, EFRI-0735974, and by the DOE under grant DE-FG52-06NA27490.

R. Fuller and X.D. Koutsoukos (Eds.): MELT 2009, LNCS 5801, pp. 66–78, 2009.
© Springer-Verlag Berlin Heidelberg 2009

improvement. Many other works followed. [5] improved upon [1] by taking the probabilistic nature of the problem into account. Another class of systems such as [6, 7] use stochastic triangulation techniques relying on some path loss model, in which the modeling error can lead to inaccuracy. In contrast, our approach is model-free and based solely on actual measurements. Our earlier related work has been shown to reduce the mean error distance by a factor of 3.5 compared to stochastic triangulation; see [8]. References to many other systems can be found in [9].

Despite the rich literature, some fundamental questions remain. The present paper not only describes a successful positioning system, but also suggests a set of formal techniques that proved to work well in the real setting. Our approach is stochastic in nature. Localization is done relative to a *landmark graph*, whose nodes are a chosen set of landmarks, and whose edges exist between any two nodes if the corresponding landmarks are in contiguous geographical areas. The device's position is mapped either to a node of the landmark graph if the device is in its vicinity, or to an edge if the device is in the area between two landmarks.

Choosing not to assume any model that describes signal propagation and postulates a way in which signal strength decreases with distance, our system is solely based on the measurements obtained at the landmarks. Naturally, we would want to use as few measurements as possible to achieve a desirable level of accuracy. In essence, our work suggests that the accuracy achieved is on the order of the landmark density and this provides a rule-of-thumb for designing a localization system. The interpolation technique between probability distributions we use in constructing location profiles is aimed at "generalizing" the discrete measurements at our disposal into descriptors that can cover a broader area and are not so sensitive to the exact position at which the measurements are taken.

As mentioned, the accuracy we achieve substantially outperforms stochastic triangulation approaches. It should be judged bearing in mind the density of the clusterheads used by our system. We note that the ratio of (possible discrete) locations to clusterheads is relatively high in our system compared to alternative approaches, implying that the deployment cost is low. Proximity-based systems, for instance, require many more clusterheads to achieve the level of accuracy we report.

Notation. We use bold lower case letters for vectors, bold upper case letters for matrices, and ^{T} denotes transpose. Our discussions will involve both probability density functions (pdfs) and probability mass functions (pmfs). With a slight abuse of terminology, we will use the term pdf throughout.

2   Problem Formulation

Consider the problem of locating a wireless sensor network device in a contiguous space $\mathcal{X}$, which typically corresponds to some indoor environment. First, we map this space with landmarks and areas connecting the landmarks — in other words, a landmark graph. Denote the set of landmarks (nodes of the graph) by $\mathcal{V} = \{V_i \mid i = 1, \ldots, M\}$, and the areas between landmarks (the edges) by $\mathcal{E} = \{E_{ij} \mid i = 1, \ldots, M, \ j > i, \ V_j \in \mathcal{N}_i\}$, where $\mathcal{N}_i$ is the set of neighboring landmarks of V_i. With a slight abuse of notation, we sometimes also write $j \in \mathcal{N}_i$ if $V_j \in \mathcal{N}_i$ and $(i, j) \in \mathcal{E}$ when $E_{ij} \in \mathcal{E}$. In what follows, a *location* refers to either a node or an edge. The set of all locations will be denoted by $\mathcal{L} = \{L_l \mid l = 1 \ldots, N\}$, where $N = M + |\mathcal{E}|$.

The next step is profiling, i.e., to associate to various locations appropriate probabilistic descriptors of some features of the wireless signal. Here we use the RSSI, which is measured between all pairs of landmarks. (Additional RF features may also be used if available.) Let $Y^{(k)} \in \{\eta_1, \ldots, \eta_H\}$ be the RSSI received at landmark k. We then have a collection of empirical distributions:

$$q_i^{(k)}(y) = \text{Freq}(Y^{(k)} = y | V_i), \ i, k = 1, \ldots, M, \tag{1}$$

where k is the index of the receiving landmark and i is the index of the transmitting landmark. Using these empirical distributions, we build the probabilistic descriptors of all locations using methods introduced in the sections that follow. As the result of profiling, we obtain a pdf of RSSI that characterizes the signals transmitted from each location and received at each landmark. In fact, for improved robustness we associate with each location a family of pdfs parametrized by vectors $\boldsymbol{\theta}_i$ and $\boldsymbol{\theta}_{ij}$, respectively. These are the location descriptors or profiles:

$$p_i^{(k)}(\cdot | \boldsymbol{\theta}_i), \ i = 1, \ldots, M, \ k = 1, \ldots, M; \tag{2}$$

$$p_{ij}^{(k)}(\cdot | \boldsymbol{\theta}_{ij}), \ (i, j) \in \mathcal{E}, \ k = 1, \ldots, M. \tag{3}$$

In the above, the pdf families listed correspond to the nodes (cf. (2)) and the edges (cf. (3)) of the landmark graph, respectively. Equivalently, we may list the pdf families in terms of the locations, with the notation

$$p_{Y^{(k)} | \boldsymbol{\theta}_l}(\cdot), \ l = 1, \ldots, N, \ k = 1, \ldots, M, \tag{4}$$

where l corresponds to a location — either a node or an edge. The former notation will be used when we discuss profiling, while the latter will be used in localization. Clusterhead placement will place $K \leq M$ clusterheads (one can think of a limited clusterhead "budget") at some of the landmarks, which will listen to the signals transmitted by the wireless device. Localization is done by "comparing" the clusterheads' RSSI measurements with the location profiles.

3   Profiling

This section focuses on how to generate the location profiles (2), (3) using the empirical RSSI distributions (1). The key technique is the interpolation of pdfs.

3.1   Interpolation of PDFs

A naive way of interpolating pdfs is to calculate a simple weighted average. However, one may quickly find that the naive way can produce unnatural results. For example, given two Gaussian pdfs with different means, their naive interpolation always has two peaks.

A more sophisticated approach has appeared in the statistical physics literature [10], which we adopt with some generalizations. Given K pdfs, $p_1(x)$, $p_2(x), \ldots, p_K(x)$, let $\mu_1, \mu_2, \ldots, \mu_K$ and $\sigma_1^2, \sigma_2^2, \ldots, \sigma_K^2$ be their means and variances, respectively. Let $\rho \in \mathbb{R}^K$ with elements $\rho_1, \rho_2, \ldots, \rho_K \in [0,1]$ satisfying $\sum_{i=1}^K \rho_i = 1$. We are now seeking an interpolation $p_\rho(x)$, whose mean and variance are $\mu_\rho = \sum_{i=1}^K \rho_i \mu_i$ and $\sigma_\rho^2 = \sum_{i=1}^K \rho_i \sigma_i{}^2$. Let

$$\xi_i(x) = \frac{\sigma_i}{\sigma_\rho}(x - \mu_\rho) + \mu_i, \quad i = 1, \ldots, l.$$

When the random variable takes discrete values, an issue is that the transformation $\xi(x)$ may produce a value for which probability is not defined. An approximate formula that solves this issue is also provided as follows. Assume that the probabilities are defined for values $-\infty, \ldots, -1, 0, 1, \ldots, \infty$. For integers j and l, and for $i = 1, \ldots, K$, let

$$\gamma_{ijl} = \max \left\{ \begin{array}{l} 0, \\ \min \left\{ \begin{array}{l} \xi_i(j + 0.5) \\ l + 0.5 \end{array} \right\} - \max \left\{ \begin{array}{l} \xi_i(j - 0.5) \\ l - 0.5 \end{array} \right\} \end{array} \right\}, \tag{5}$$

The interpolation formula is then

$$p_\rho(j) = \sum_{i=1}^K \rho_i \sum_l \gamma_{ijl} \cdot p_i(l). \tag{6}$$

We call this formula the *linear interpolation*. Similarly to [10], one can prove that μ_ρ and σ_ρ^2 are indeed the mean and variance corresponding to $p_\rho(x)$. From here on, we denote the linear interpolation of the K pdfs with the coefficient vector $\rho \in \mathbb{R}^K$ by $\mathrm{Interpol}(\rho, p_1, p_2, \ldots, p_K)$.

3.2   Associating PDF Families to Locations

It suffices to consider the RSSI profile of all locations observed by a clusterhead placed at one of the landmarks. The index of the clusterhead is thus suppressed in all formulae of this subsection.

First, we "regularize" the empirical pdfs to get rid of zero elements. This is necessary because the size of our sample during profiling is finite. As a result, some RSSI value η_h that is possible but rare for a location L_i might not be observed during profiling, leaving the hth element of the empirical pdf equal to zero. If we use the empirical pdf directly as the probabilistic descriptor of the location, then when η_h appears, we would rule out L_i immediately, regardless of how many total observations are made and how the rest of the observations resemble the profile of location L_i. This is clearly undesirable. To solve this problem, we mix the empirical pdf with a discretized Gaussian-like pdf of the same mean and variance. Namely, let q be an empirical pdf with mean μ and variance σ^2. Let $\phi(\mu, \sigma^2)$ be a Gaussian-like pdf whose domain is discretized to the set $\{\eta_1, \ldots, \eta_H\}$. Let $\gamma \in (0,1)$ be a chosen mixing factor — typically we set γ to a small value such as 0.1 or 0.2. Then the pdf after regularization is $\tilde{q} = (1 - \gamma)q + \gamma\phi(\mu, \sigma^2)$.

Second, the landmarks are characterized by pdf-families constructed using interpolation. See [11] for more discussion on the robustness of the pdf-family framework. However, [11] did not provide a formal technique for constructing the pdf-families. Specifically, suppose V_i has I neighbors — $\mathcal{N}_i = \{V_{j_1}, \ldots, V_{j_I}\}$. Let $\boldsymbol{\rho}_{\boldsymbol{\theta}_i} = (1 - \sum_{j=1}^{I} \theta_i^{(j)}, \ \theta_i^{(1)}, \ldots, \theta_i^{(I)})^T$, where $\boldsymbol{\theta}_i \in \mathbb{R}^I$, $\boldsymbol{\theta}_i \geq \mathbf{0}$ elementwise, and $\sum_{j=1}^{I} \theta_i^{(j)} < 1$. Then the pdf family associated with V_i can be defined as an interpolation of $I + 1$ empirical pdfs:

$$p_i(\cdot|\boldsymbol{\theta}_i) \triangleq \text{Interpol}\left(\boldsymbol{\rho}_{\boldsymbol{\theta}_i}, \tilde{q}_i(\cdot), \tilde{q}_{j_1}(\cdot), \ldots, \tilde{q}_{j_I}(\cdot)\right).$$

Last, consider the edges of the landmark graph. As will be justified by the experiments, we associate with the edge (i, j) a pdf family defined as the interpolation of the pdf families for landmarks i and j. Let $\vartheta_{ij} \in (0,1)$ and $\boldsymbol{\theta}_{ij}$ be a vector concatenating $\boldsymbol{\theta}_i$, $\boldsymbol{\theta}_j$, and ϑ_{ij}. The pdf family associated with edge (i, j) is

$$p_{ij}(\cdot|\boldsymbol{\theta}_{ij}) \triangleq \text{Interpol}\left(\begin{bmatrix} \vartheta_{ij} \\ 1 - \vartheta_{ij} \end{bmatrix}, p_i(\cdot|\boldsymbol{\theta}_i), p_j(\cdot|\boldsymbol{\theta}_j)\right).$$

3.3   An Alternative Gaussian Model

In the above, we focused on associating a family of generally shaped pdfs to each location. If a Gaussian model of the RSSI is used instead, this task can be greatly simplified. One may then ask whether using generally shaped pdfs is worth the effort. The answer to this question may depend on circumstances. However, our experiments show that significant information regarding the signals transmitted from a location is captured by our approach, but would be neglected if we assume the Gaussian model.

4   Localization System Design

4.1   Binary Composite Hypothesis Testing

We start our analysis by considering the simpler problem of using a single clusterhead at landmark V_k to localize a device whose location is either L_i or L_j. The pdf families associated with the two locations are $p_{\mathbf{Y}^{(k)}|\boldsymbol{\theta}_i}(\mathbf{y})$ and $p_{\mathbf{Y}^{(k)}|\boldsymbol{\theta}_j}(\mathbf{y})$, respectively. The clusterhead makes n i.i.d. observations $\mathbf{y}^{(k),n} = (\mathbf{y}_1^{(k)}, \ldots, \mathbf{y}_n^{(k)})$.

The likelihood of obtaining these measurements if L_i is the true location is
$p_{\mathbf{Y}^{(k)}|\boldsymbol{\theta}_i}(\mathbf{y}^{(k),n}) = \prod_{l=1}^{n} p_{\mathbf{Y}^{(k)}|\boldsymbol{\theta}_i}(\mathbf{y}_l^{(k)})$.

The problem at hand is a binary composite hypothesis testing problem for which the *Generalized Likelihood Ratio Test (GLRT)* is commonly used. The GLRT compares the normalized generalized log-likelihood ratio

$$X_{ijk}(\mathbf{y}^{(k),n}) = \frac{1}{n} \log \frac{\sup_{\boldsymbol{\theta}_i \in \Omega_i} p_{\mathbf{Y}^{(k)}|\boldsymbol{\theta}_i}(\mathbf{y}^{(k),n})}{\sup_{\boldsymbol{\theta}_j \in \Omega_j} p_{\mathbf{Y}^{(k)}|\boldsymbol{\theta}_j}(\mathbf{y}^{(k),n})}$$

to a threshold λ, and declares L_i whenever

$$\mathbf{y}^{(k),n} \in \mathscr{S}_{ijk,n}^{GLRT} \triangleq \{\mathbf{y}^n \mid X_{ijk}(\mathbf{y}^n) \geq \lambda\},$$

and L_j otherwise. There are two types of error (referred to as type I and type II, respectively) with probabilities

$$\alpha_{ijk,n}^{GLRT}(\boldsymbol{\theta}_j) = \mathbf{P}_{\boldsymbol{\theta}_j}[\mathbf{y}^{(k),n} \in \mathscr{S}_{ijk,n}^{GLRT}],$$

$$\beta_{ijk,n}^{GLRT}(\boldsymbol{\theta}_i) = \mathbf{P}_{\boldsymbol{\theta}_i}[\mathbf{y}^{(k),n} \notin \mathscr{S}_{ijk,n}^{GLRT}],$$

where $\mathbf{P}_{\boldsymbol{\theta}_j}[\cdot]$ (resp. $\mathbf{P}_{\boldsymbol{\theta}_i}[\cdot]$) is a probability evaluated assuming that L_j (resp. L_i) is the true location. We will use the term *exponent* to refer to the quantity $\lim_{n\to\infty} \frac{1}{n} \log \mathbf{P}[\cdot]$ for some probability $\mathbf{P}[\cdot]$; if the exponent is d then the probabilities approaches zero as e^{-nd}.

For any sequence of observations $\mathbf{y}^n = (\mathbf{y}_1, \ldots, \mathbf{y}_n)$, the empirical measure (or type) is given by $\mathbf{L}_{\mathbf{y}^n} = (L_{\mathbf{y}^n}(\sigma_1), \ldots, L_{\mathbf{y}^n}(\sigma_{|\Sigma|}))$, where

$$L_{\mathbf{y}^n}(\sigma_i) = \frac{1}{n} \sum_{j=1}^{n} \mathbf{1}\{\mathbf{y}_j = \sigma_i\}, \qquad i = 1, \ldots, |\Sigma|,$$

and $\mathbf{1}\{\cdot\}$ denotes the indicator function. We will denote the set of all possible types of sequences of length n by $\mathscr{L}_n = \{\boldsymbol{\nu} \mid \boldsymbol{\nu} = \mathbf{L}_{\mathbf{y}^n} \text{ for some } \mathbf{y}^n\}$ and the type class of a probability law $\boldsymbol{\nu}$ by $T_n(\boldsymbol{\nu}) = \{\mathbf{y}^n \in \Sigma^n \mid \mathbf{L}_{\mathbf{y}^n} = \boldsymbol{\nu}\}$. Last, recall that the Kullback-Leibler (KL) distance of $\boldsymbol{\nu}$ from another pdf $\boldsymbol{\mu}$ is

$$D(\boldsymbol{\nu}\|\boldsymbol{\mu}) = \sum_{i=1}^{|\Sigma|} \boldsymbol{\nu}(\sigma_i) \log \frac{\boldsymbol{\nu}(\sigma_i)}{\boldsymbol{\mu}(\sigma_i)}. \tag{7}$$

In our previous work [8] we derived bounds on the type I and type II error probability exponents:

$$\limsup_{n\to\infty} \frac{1}{n} \log \alpha_{ijk,n}^{GLRT}(\boldsymbol{\theta}_j) \leq -\lambda, \tag{8}$$

$$\limsup_{n\to\infty} \frac{1}{n} \log \beta_{ijk,n}^{GLRT}(\boldsymbol{\theta}_i) \leq -\inf_{\mathbf{Q}\in\mathscr{D}_{ijk}} D(\mathbf{Q}\|\mathbf{P}_{\boldsymbol{\theta}_i}), \tag{9}$$

for all $\boldsymbol{\theta}_j \in \Omega_j$ and $\boldsymbol{\theta}_i \in \Omega_i$, where

$$\mathscr{D}_{ijk} = \{\mathbf{Q}| \inf_{\boldsymbol{\theta}_j} D(\mathbf{Q}\|\mathbf{P}_{\boldsymbol{\theta}_j}) - \inf_{\boldsymbol{\theta}_i} D(\mathbf{Q}\|\mathbf{P}_{\boldsymbol{\theta}_i}) < \lambda\}.$$

4.2   Determining the Optimal Threshold

It can be seen from (8) and (9) that the exponent of the type I error probability is increasing with λ but the exponent of the type II error probability is nonincreasing with λ. We have no preference between the two types of error, thus, we wish to balance the two exponents and determine the value of λ at which they become equal.

The exponent of the type I error is simply obtained from (8). The type II error exponent from (9) is equivalent to

$$Z_{ijk}(\lambda, \boldsymbol{\theta}_i) = \min_{\mathbf{Q}} D(\mathbf{Q}\|\mathbf{P}_{\boldsymbol{\theta}_i}) \\ \text{s.t.} \ \min_{\boldsymbol{\theta}_j} D(\mathbf{Q}\|\mathbf{P}_{\boldsymbol{\theta}_j}) - D(\mathbf{Q}\|\mathbf{P}_{\boldsymbol{\theta}_i}) \leq \lambda, \ \forall \boldsymbol{\theta}_i. \tag{10}$$

The worst case exponent over $\boldsymbol{\theta}_i \in \Omega_i$ is given by

$$Z_{ijk}(\lambda) = \min_{\boldsymbol{\theta}_i} Z_{ijk}(\lambda, \boldsymbol{\theta}_i).$$

Problem (10) is nonconvex; we use dual relaxation to obtain a quantity that is easier to compute. Let $\bar{Z}_{ijk}(\lambda, \boldsymbol{\theta}_i)$ be the optimal value of the dual of (10); by weak duality it follows $Z_{ijk}(\lambda, \boldsymbol{\theta}_i) \geq \bar{Z}_{ijk}(\lambda, \boldsymbol{\theta}_i)$. It can be verified that there exists a $\lambda_{ijk}^* > 0$ such that $\bar{Z}_{ijk}(\lambda_{ijk}^*) = \lambda_{ijk}^*$. Furthermore, both error exponents in (8) and (9) are no smaller than λ_{ijk}^*.

Now suppose the clusterhead at V_k has obtained the measurements $\mathbf{y}^{(k),n}$ and seeks to decide the device location between L_i and L_j. The clusterhead has the option of using the GLRT by comparing $X_{ijk}(\mathbf{y}^{(k),n})$ to the threshold λ_{ijk}^*, or comparing $X_{jik}(\mathbf{y}^{(k),n})$ to a threshold λ_{jik}^* that can be obtained in exactly the same way as λ_{ijk}^*. We thus let

$$d_{ijk} = \max\{\lambda_{ijk}^*, \lambda_{jik}^*\}, \tag{11}$$

and set $(\bar{i}, \bar{j}) = (i, j)$ if λ_{ijk}^* is the maximizer above; otherwise set $(\bar{i}, \bar{j}) = (j, i)$. Define the maximum probability of error as

$$P_{ijk,n}^{(e)} \triangleq \max\{\max_{\boldsymbol{\theta}_{\bar{j}}} \alpha_{ijk,n}^{GLRT}(\boldsymbol{\theta}_{\bar{j}}), \max_{\boldsymbol{\theta}_{\bar{i}}} \beta_{ijk,n}^{GLRT}(\boldsymbol{\theta}_{\bar{i}})\}.$$

The following result provides a performance guarantee.

Proposition 1. *Suppose that the clusterhead at V_k uses the GLRT and compares $X_{\bar{i}\bar{j}k}(\mathbf{y}^{(k),n})$ to d_{ijk}. Then, the maximum probability of error satisfies*

$$\limsup_{n\to\infty} \frac{1}{n} \log P_{ijk,n}^{(e)} \leq -d_{ijk}.$$

4.3   Multiple Composite Hypothesis Testing

We assume without loss of generality that the clusterheads $1, 2, \ldots, K$ are placed at positions $V_1, V_2, \ldots, V_K$. Let d_{ijk} be the GLRT threshold obtained in Sec. 4.2 for each location pair (i, j), $i < j$, and clusterhead k.

We make $N-1$ binary decisions with the GLRT. Specifically, we first compare L_1 with L_2 to accept one hypothesis, then compare the accepted hypothesis with L_3, so on and so forth. For each one of these L_i vs. L_j decisions we use a single clusterhead k as detailed in Sec. 4.1, hence the exponent of error probability is bounded by d_{ijk}. All in all we make $N-1$ binary hypothesis tests, each involving a single (potentially different) clusterhead. These clusterheads can collaborate in a distributed fashion as we have shown in [11] to make the final decision.

4.4 Clusterhead Placement

Being able to optimize clusterhead placement is one important benefit of our hypothesis testing approach, which produced error bounds that can serve as the criterion. Consider an arbitrary placement of K clusterheads. More specifically, let $\mathscr{Y}$ be any subset of the set of potential clusterhead positions $\mathscr{B}$ with cardinality K. Let $\mathbf{x}(\mathscr{Y}) = (x_1(\mathscr{Y}), \ldots, x_M(\mathscr{Y}))$ where $x_k(\mathscr{Y})$ is the indicator function of B_k being in $\mathscr{Y}$. The objective of clusterhead placement problem is to minimize the worst case probability of localization error, that is, to find ϵ^* as

$$\epsilon^* = \max_{\mathscr{Y}} \min_{\substack{i,j=1,\ldots,N \\ i<j}} \max_{k:x_k(\mathscr{Y})=1} d_{ijk}. \tag{12}$$

This combinatorial optimization problem can be rewritten as a mixed integer linear programming problem (MILP). Although it is NP-hard, it can be solved efficiently by using a special purpose algorithm from [12].

We will use the decision rule outlined in Section 4.3 and for every region pair (i,j) we will rely on the clusterhead at $B_{k_{ij}^*}$ to make the corresponding decision. The following theorem establishes a performance guarantee.

Proposition 2. *Place clusterheads according $\mathscr{Y}^* \triangleq \{B_k | x_k^* = 1\}$ and for every (i,j) select one clusterhead with index k_{ij}^* so that $d_{ijk_{ij}^*} = \max_{k:x_k(\mathscr{Y})=1} d_{ijk}$. Then, the worst case probability of error for the decision rule described in Section 4.3, $P_n^{(e),opt}$, satisfies*

$$\limsup_{n \to \infty} \frac{1}{n} \log P_n^{(e),opt} \leq -\epsilon^*. \tag{13}$$

5 Experiments

Our testbed is set up on the first floor of a Boston University building (see Fig. 4), and uses MPR2400 (MICAz) motes from Crossbow Technology Inc.

5.1 Testing PDF Interpolations

We have proposed a rather sophisticated interpolation technique for generating location profiles. One concern is: if the interpolated pdfs were merely low-quality approximations of the actual pdfs, then we might be better off using a Gaussian approximation, which is computationally cheaper. In our experiments however, the interpolated pdfs did a very good job preserving the information that resides in the shapes of the empirical pdfs. As will be shown, the decision accuracy using the interpolated pdfs dominates that of the Gaussian approximation by a significant margin. Another question that we attempt to answer is: At what length scale does pdf interpolation make sense? It turns out that the interpolation is very meaningful when the two end points are about 30 feet (or 9 meters) apart, but not when they are 75 feet apart.

Ideally, and in order to improve accuracy, one would like to place landmarks as close as possible implying that we would need to interpolate between points that are 30 feet (or less) apart. However, it turns out that interpolations over points that are less than 30 feet apart may not be worth the effort. This is consistent with results reported in [2], which have shown that when the spacing of "reference signatures" goes below roughly 10 meters, the improvement in performance diminishes. (The spacing of the "reference signatures" is analogous to the distance between the two end-point locations in our pdf interpolation.) This result reinforces that of [2], as both indicate that taking empirical measurements at a spacial density of less than 9 or 10 meters apart, or roughly 1 per 25 sq. meters, carries diminishing benefit.

This experiment is conducted in a corridor of roughly 75 feet long, mapped to 6 locations roughly 15 feet apart. A clusterhead (the receiver) is placed at location 1. To measure the signals transmitted from each location, one of the coauthors stood at that location holding a transmitting mote, which sends a packet every 5 seconds. We chose to have a person hold the mote because this is close to an actual application scenario. The clusterhead received the packets and recorded the RSSI values. During the experiment, a total of 150 packets were sent from each location. Due to packet loss, the number of actual samples taken by the clusterhead is less, but we still obtained more than 100 samples for each location. Then, we mix a Gaussian component into each of the six empirical distributions as described earlier with a mixing factor of 0.2, i.e., regularized empirical distribution = 0.8 measured + 0.2 Gaussian. The empirical distributions for the six locations after regularization are denoted by $q_1, q_2, \ldots, q_6$.

We compare three interpolation methods. First, in what is labeled "short interpolation", the interpolated pdf of location i is generated using q_{i+1} and q_{i-1}:

$$p_{i,\text{short}} = \text{Interpol}\left(\begin{bmatrix} 0.5 \\ 0.5 \end{bmatrix}, q_{i-1}, q_{i+1} \right), \; i = 2, 3, 4, 5.$$

Second, in what is labeled "long interpolation", the interpolated pdfs are generated using q_1 and q_6:

$$p_{i,\text{long}} = \text{Interpol}\left(\begin{bmatrix} \frac{6-i}{5} \\ \frac{i-1}{5} \end{bmatrix}, q_1, q_6 \right), \quad i = 2, 3, 4, 5.$$

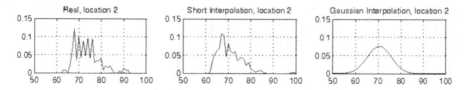

Fig. 1. Visual comparison of interpolated pdfs for location 2. From left to right: empirical pdf, short (linear) interpolation, and Gaussian approximation.

Third, we adopt the Gaussian model instead and interpolate the pdf of each location with adjacent locations:

$$p_{i,\text{gaussian}} = \text{Interpol}\left(\begin{bmatrix} 0.5 \\ 0.5 \end{bmatrix}, \phi(\mu_{i-1}, \sigma^2_{i-1}), \phi(\mu_{i+1}, \sigma^2_{i+1})\right)$$
$$= \phi\left(\frac{\mu_{i-1}+\mu_{i+1}}{2}, \frac{\sigma^2_{i-1}+\sigma^2_{i+1}}{2}\right), \qquad i = 2, 3, 4, 5.$$

Qualitative Study. In the interest of space, we only visually compare the short interpolation and the Gaussian approximation for location 2 in Fig. 1.

The short interpolation seems to capture some shape information of the actual pdf that is missed by the Gaussian model. For example, the empirical pdf is skewed to the left. The interpolated pdf also exhibits the skewness, while the Gaussian pdf is always symmetrical.

Quantitative Study. First, it is of interest to compare the qualities of the different interpolations using the Kullback-Leibler (K-L) distance (cf. Eq. (7)) as a metric of distance between pdfs. This information theoretic distance is closely related to statistics, including the results derived in the present paper; see Section 4.

The comparison is plotted in Fig. 2. It is very interesting to see that *the quality of short interpolation dominates that of the Gaussian model*. For example, the K-L distance of short-interpolation-to-empirical for location 4 is only a little over one third of that of the Gaussian model. For locations 2 and 3, the difference is roughly a factor of 1.5, which is still significant. The long interpolation on the other hand clearly departs significantly from the actual distribution.

5.2 Testing the Complete System

Our localization system covers 10 rooms and the corridors, which are mapped to 30 landmarks, marked by either a green circle or a red square on the floor plan (Fig. 4). The landmark graph is then constructed resulting in 39 edges, or a total of 69 locations. Hence $N = 69$, $M = 30$ and $1 \leq K \leq 30$ in this experiment. A mote is placed at the center of each landmark location, but only some of them will serve as clusterheads. All 30 motes are connected to a base MICAz through a mesh network. The base mote is docked on a Stargate node which forwards the beacon message back to server.

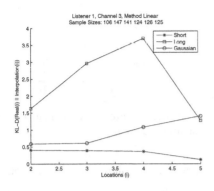

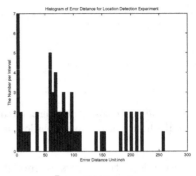

$$\bar{D}_e = 87.32 \text{ inches}$$

Fig. 2. Kullback-Leibler distance comparison

Fig. 3. Localization result

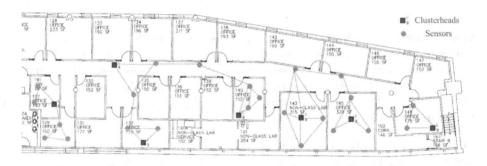

Fig. 4. Floor plan with the landmarks for the testbed

The experimental validation of our localization approach can be divided into the five phases:

Phase 1. We obtained the empirical pdfs for the landmarks corresponding to Eq. (1). With 30 motes placed at each landmark location, we scheduled them such that the motes took turns to broadcast packets, and when one was transmitting the others would listen and record the RSSI. A total of 200 packets were transmitted by each mote. The data collection was repeated for the combinations of two frequencies and two power levels; details will be given below.

Phase 2. We used the methods in Section 3 to construct the pdf families corresponding to Eqs. (2), (3) which are the descriptors of all 69 locations — both the landmarks and the edges of the landmark graph. Note that the interpolation technique allowed us to construct high quality descriptors without densely covering the area while collecting empirical measurements.

Phase 3. We obtained d_{ijk} as described in Sec. 4.

Phase 4. We solved the MILP to optimize clusterhead placement and simultaneously obtained the performance guarantee (Prop. 2). In the MILP formulation, we needed to input K, the total number of clusterheads. By varying K from 1 to 30, we discovered that the performance guarantee reached a satisfactory level after $K = 7$, and somewhat flattens afterward. Thus, we assigned 7 of the 30 motes as clusterheads. (Note the low clusterhead density needed by our system). The optimal placement is marked in Fig. 4 by the red squares.

Phase 5. We placed motes in the coverage area, let them broadcast messages, then possibly move some of them and let them broadcast again, and finally let the clusterheads report their localization and movement detection decisions.

We let Phase 1 (a completely automated procedure) stretch over 24 hours to acquire data under diverse conditions of the surrounding environment. Phase 2 takes virtually no time. Phase 3 is the most time-consuming part and takes another 24 hours on our computer, although further optimization of our code may reduce the computation time significantly. Phase 4 only takes about half an hour. All these steps only need to be done once.

We know from previous experiences that frequency and power diversity provide better performance [8]. We made 56 localization tests in random positions throughout the covered area. The mote to be located broadcasted 20 packets over the combination of 2 frequencies (2.410 GHz and 2.460 GHz) and 2 power levels (0 dBm and -10 dBm), with 5 packets for each combination. We achieved a mean error distance of 87.32 inches, which is better than our earlier result of 96.08 inches [8] based on techniques that do not use a formal method of pdf interpolation. The percentile of errors< 3 meters (118 inches) also improved from 80% to 87%. One may also count from Fig. 3 that the percentile of errors< 5 meters (197 inches) is 95%.

The total coverage area (we have excluded the rooms that are in the floor plan but to which we do not have access thus have not placed a mote) was 1827 feet^2, that is, about 61 feet^2 per landmark. With a mean error distance of $\bar{D}_e = 7.3$ feet the mean area of "confusion" was $7.3^2 = 53$ feet^2. It is evident that we were able to achieve accuracy on the same order of magnitude as the area "covered" by a landmark; this is the best possible outcome with a "discretized" system such as ours. That is, the system was identifying the correct location or a neighboring location most of the time. We used a clusterhead density of 1 clusterhead per $1827/7 = 261$ feet^2. Note that our system is *not* based on the "proximity" to a clusterhead; the ratio of locations to clusterheads is 69/7, or about 10.

6   Conclusion

The paper reports a landmark-based localization system where each hypothesis is associated with a family of pdfs constructed by a pdf linear interpolation technique. Both theoretical and experimental justifications are provided.

References

1. Bahl, P., Padmanabhan, V.: RADAR: An in-building RF-based user location and tracking system. In: Proceedings of the IEEE INFOCOM Conference, Tel-Aviv, Israel (2000)
2. Lorincz, K., Welsh, M.: Motetrack: A robust, decentralized approach to RF-based location tracking. In: Springer Personal and Ubiquitous Computing, Special Issue on Location and Context-Awareness, pp. 1617–4909 (2006)
3. Kaemarungsi, K., Krishnamurthy, P.: Modeling of indoor positioning systems based on location fingerprinting. In: Proceedings of the IEEE INFOCOM Conference (2004)
4. Hightower, J., Want, R., Borriello, G.: SpotON: An indoor 3d location sensing technology based on RF signal strength. UW CSE 00-02-02, University of Washington, Department of Computer Science and Engineering, Seattle, WA (February 2000)
5. Castro, P., Chiu, P., Kremenek, T., Muntz, R.: A probabilistic location service for wireless network environments. In: Proceedings of Ubicomp, Atlanta, GA. ACM, New York (2001)
6. Patwari, N., Hero, A.O., Perkins, M., Correal, N.S., O'Dea, R.J.: Relative location estimation in wireless sensor networks. IEEE Transanctions on Signal Processing 51(8), 2137–2148 (2003)
7. Yedavalli, K., Krishnamachari, B., Ravula, S., Srinivasan, B.: Ecolocation: A sequence based technique for rf-only localization in wireless sensor networks. In: The Fourth International Conference on Information Processing in Sensor Networks, Los Angeles, CA (April 2005)
8. Paschalidis, I.C., Guo, D.: Robust and distributed stochastic localization in sensor networks: Theory and experimental results. ACM Transactions on Sensor Network 5(4) (November 2009)
9. Youssef, M.A.: Collection about location determination papers (2008), http://www.cs.umd.edu/~moustafa/location_papers.htm
10. Bursal, F.H.: On interpolating between probability distributions. Applied Mathematics and Computation 77, 213–244 (1996)
11. Paschalidis, I.C., Guo, D.: Robust and distributed localization in sensor networks. In: Proceedings of the 46th IEEE Conference on Decision and Control, New Orleans, Louisiana, December 2007, pp. 933–938 (2007)
12. Ray, S., Lai, W., Paschalidis, I.C.: Statistical location detection with sensor networks. Joint special issue IEEE/ACM Trans. Networking and IEEE Trans. Information Theory 52(6), 2670–2683 (2006)

A Calibration-Free Localization Solution for Handling Signal Strength Variance

Fangfang Dong, Yiqiang Chen, Junfa Liu, Qiong Ning, and Songmei Piao

Institute of Computing Technology, Chinese Academy of Sciences, Beijing, China
{dongfangfang,yqchen,liujunfa,ningqiong,piaosongmei}@ict.ac.cn

Abstract. In RSS-based indoor localization techniques, signal strength variance between diverse devices can significantly degrade the positional accuracy when using the radio map derived by train device to other test device. Current solutions employ extra calibration data from test device to solve this problem. In this paper, we present a calibration-free solution for handling the signal strength variance between diverse devices. The key idea is to generate radio map using signal strength differences between pairs of APs instead of absolute signal strength values. The proposed solution has been evaluated by extending with two well-known localization technologies. We evaluate our solution in a real-world indoor wireless environment and the results show that the proposed solution solves the signal strength variance problem without extra calibration on test device and performs equally to that of existing calibration-based method.

Keywords: Calibration-Free Localization, Signal Strength Variance, Wireless Local-Area Network.

1 Introduction

Location estimation is an important prerequisite for many awareness applications in ubiquitous computing. In an indoor environment, increasing attention is paid to localization using the popular and inexpensive 802.11 Wireless Local-Area Network (WLAN) as the fundamental infrastructure. In general, WLAN based localization methods work in two phases: an off-line training phase and an online localization phase. In the off-line phase, human carrying a mobile device needs to walk around the interest area and collect signal strength values received from various access points (APs). These values comprise a radio map of the physical region, which is compiled into a deterministic or statistical prediction model for the online phase. In the online phase, the real-time signal strength samples are used to lookup in the radio map to estimate the current location based on the learned model.

A fundamental problem cannot be ignored is the signal strength variance between diverse devices. Due to lack of standardization and inequalities in hardware and software, various mobile devices have apparently distinct capacities of sensing wireless signals. Therefore, the distributions of received signal strength

R. Fuller and X.D. Koutsoukos (Eds.): MELT 2009, LNCS 5801, pp. 79–90, 2009.
© Springer-Verlag Berlin Heidelberg 2009

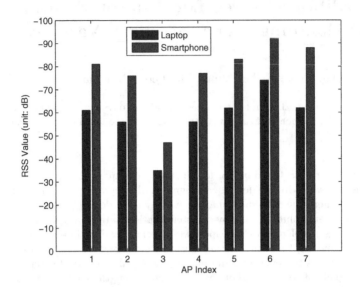

Fig. 1. Signal strength detected by an IBM laptop and an O2 smartphone at a fixed location

(RSS) values collected by different devices also vary with each other. Figure 1 shows the discrepancy of signal distribution caused by different devices. Here we use an IBM R60 laptop and an O2 Xda Atom Life smartphone to collect RSS for several seconds at a fixed location separately. Then we average those measurements on each device to get the results.

From Figure 1 we can inference that if the radio map derived by train device is directly used on other test device, the localization accuracy will drop down greatly. As reported in [1], the room-size localization accuracy drops to unusable 10% under this kind of situation. Current solutions mainly employ some extra manual measurements on test device to find mappings between train and test devices [2] [3] [4] [5]. However, these methods are very limited because it's a time consuming work to collect measurements on test devices. Furthermore, considering the huge number of different IEEE 802.11 clients on the market, it's unpractical to do this work for each kind of devices.

In this paper, we propose a calibration-free solution to solve the signal strength variance problem between diverse devices, which we called DIFF. The key idea of DIFF is to use signal strength differences between pairs of APs instead of absolute signal strength values. In the off-line phase, the radio map is built by signal strength differences between pairs of APs extracted from RSS collections on train device. In the online phase, test device's location can be estimated by comparing the signal strength differences between pairs of APs extracted from real-time collections with entries of radio map. The truth behind DIFF is that although different devices receive apparently distinct signal strength values in the same location, they reflect the same relationship indicating the distance to

APs: strongest RSS from the nearest AP and weakest RSS from the farthest AP. We can observe that the signal strength patterns are almost the same for different devices at a fixed location. The pattern can be expressed by the signal strength differences between pairs of APs. The advantage of DIFF is that it could make the radio map capable for diverse devices with no extra calibration effort. [1] proposes a similar method named HLF, which uses signal strength ratios between pairs of APs instead of absolute signal strength values. However, after analysis of the RSS data we find difference feature of DIFF is more reasonable than ratio feature used in HLF. We have evaluated DIFF by combining with two well-known localization technologies: Nearest Neighbor and Bayesian Inference. They have been tested on data set collected with two different mobile devices in a real-world environment. The results show that DIFF performs better than HLF and is equal to calibration-based methods.

The contributions of our work are as follows: We analyse RSS values of diverse devices and show that signal strength difference feature is more reliable than ratio feature or RSS value. We extend DIFF with two typical localization technologies and show that they perform better than HLF and are equal to calibration-based methods.

The rest of this paper is organized as follows. In Section 2, we survey related works of dealing with diverse devices. We present our methodology in detail in Section 3. The experiment results and discussion are reported in Section 4. Finally, we present our conclusions and future works in Section 5.

2   Related Work

Most machine-learning based indoor localization methods are based on the radio map techniques, which can be classified into two categories: deterministic techniques and probabilistic techniques. Deterministic techniques [6] [7] [8] use deterministic inference methods to estimate a user's location, such as Triangulation and K Nearest Neighbor (KNN) used in RADAR system [6]. Probabilistic techniques [9] [10] [11] [12] [13] construct the signal strength distributions over different locations in the radio map and use probabilistic inference methods for localization, such as Horus system [9]. These traditional techniques are only limited to single device and do not address the problem of diverse devices.

[2] treats signal variation as a Gaussian mean-value shift and uses a liner model to fit the RSS values on train and test devices. [4] shows that the simple adaptation method does not work well in a complex indoor environment. Instead, it considers the transfer learning problem over devices and treats multiple devices as multiple learning tasks. Then it learns the classifier in a latent feature space. [5] utilizes train device's data and test device's partial data as benchmarks to learn a corresponding relationship in a low-dimensional space using Manifold Alignment. Then the relationship is used to transfer knowledge from train domain to help the classification in test domain. However, both of these solutions are based on manual measurements on test device to find mappings between train and test devices. Considering the huge amount of different IEEE 802.11

clients on the market, these methods are unpractical to use. [2] [3] and [14] also propose solutions that avoid manual measurement collection by learning from online-collected measurements. However, both of these solutions require a learning period and they perform considerably worse in terms of accuracy than the manual solutions.

[1] proposes a calibration-free solution for handling signal strength variance, which records fingerprints as signal strength ratios between pairs of APs instead of absolute signal strength values. While their work and ours are similar in some ways, they also differ in significant ways: (1) they use signal strength ratios between pairs of APs while we observe that signal strength differences between pairs of APs are more reliable feature. (2) the experiment results show that our method performs better than HLF used in [1]. (3) computing cost of extracting difference features are lower than that of ratio features.

3   Methodology

In this section, we first introduce the basic idea of using signal strength differences between pairs of APs. Then extend it with two typical machine-learning based localization methods.

3.1   Signal Strength Differences

Due to lack of standardization and inequalities in hardware and software, various mobile devices have apparently distinct capacities of sensing wireless signals. Therefore, the RSS values collected by different devices vary a lot with each other. Viewing Figure 1 in another way, we can get Figure 2.

We can observe from Figure 2 that although the RSS values of the two devices have about $20dB$ difference, the signal patterns composed by RSS values are very similar. In other words, the shapes of these two curves are almost the same. This is because at a fixed location, different RSS values of diverse devices reflect the same relationship indicating their distance to APs: in simple environment, strongest RSS from the nearest AP and weakest RSS from the farthest AP. The curve's shape can be expressed by the signal strength differences between pairs of APs. Therefore, we can extract pairwise differences of RSS values to replace raw RSS values.

In the off-line phase, the radio map is derived by difference features extracted from train device's RSS. In the online phase, test device's location can be estimated by comparing the signal strength difference features extracted from real-time RSS collections with entries of radio map. The advantage of this method is that it could make the radio map capable for diverse devices without any extra calibration effort.

Let's define the problem formally. Suppose there are totally m APs $B = \{b_1, b_2, ...b_m\}$ deployed in a two-dimensional wireless environment $C \in \mathbb{R}^2$. We model the physical area of interest as a finite location-state space $L = \{l_1, l_2, ...l_n\}$.

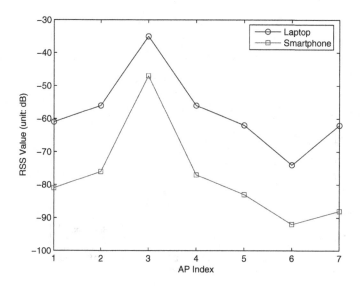

Fig. 2. Signal strength detected by an IBM laptop and an O2 smartphone at a fixed location

The state space L is denoted as a set of physical locations with x- and y- coordinates:

$$L = \{l_1 = (x_1, y_1), l_2 = (x_2, y_2), ..., l_n = (x_n, y_n)\} \tag{1}$$

As an example, each location l represents a grid cell in the environment. A mobile device taken by a user can receive wireless signals from APs periodically. The RSS values can be defined as a row vector $s = (s_1, s_2, ..., s_m) \in \mathbb{R}^m$, where s_i stands for the RSS value received from b_i. All possible signal strength values are modeled as a finite observation space $O = \{o_1, o_2, ..., o_r\}$. An observation o in the observation space O consists of a set of m signal strength measurements received from m APs. Thus, each observation o is represented as a vector of m pairs as follows:

$$o = \{(b_1, s_1), (b_2, s_2), ..., (b_m, s_m)\} \tag{2}$$

where b_k represents the kth AP scanned and s_k is the signal strength received from b_k.

The signal strength difference d is defined for a unique AP pair $b_i \times b_j \in B \times B$ with the constrain $i < j$ for uniqueness. The signal strength difference d can be computed from two observations $o_i = (b_i, s_i) \in O$ and $o_j = (b_j, s_j) \in O$ as follows:

$$d(b_i, b_j) = s_i - s_j \quad 1 \le i < j \le m \tag{3}$$

Thus, the signal strength difference feature vector D extracted from s can be expressed as follows:

$$D = (d(b_1, b_2), d(b_1, b_3), ..., d(b_{m-1}, b_m)) \tag{4}$$

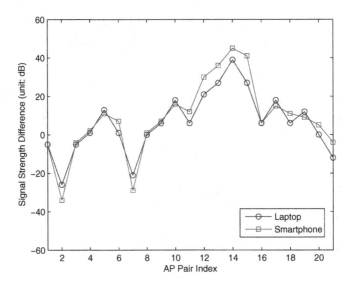

Fig. 3. Signal strength difference extracted from RSS

where the length of D is t and $t = C_m^2$.

For example, in figure 1,

$$s_{laptop} = (-61, -56, -35, -56, -62, -74, -62)$$

$$s_{smartphone} = (-81, -76, -47, -77, -83, -92, -88)$$

then

$$D_{laptop} = (-5, -26, -5, 1, 13, 1, -21, 0, 6, 18, 6, 21, 27, 39, 27, 6, 18, 6, 12, 0, -12)$$

$$D_{smartphone} = (-5, -34, -4, 2, 11, 7, -29, 1, 7, 16, 12, 30, 36, 45, 41, 6, 15, 11, 9, 5, -4)$$

as shown in figure 3. We can observe from figure 3 that the extracted difference features of two diverse devices are much more similar than RSS values.

3.2 Extended Localization Methods

In this section we present the extension of DIFF with two typical localization methods: Nearest Neighbor (NN) and Bayesian Inference (BI). The main change is replacement of absolute signal strength values with signal strength difference features.

Nearest Neighbor. As a deterministic technique to estimate a user's location, NN is first used in RADAR. It maintains a radio map from off-line RSS collections, then with which each online signal strength measurement is compared.

The coordinates of the best match location are used to give estimation. Traditional NN technique computes Euclidean distance in RSS space to find the nearest neighbor, while our method DIFF-NN needs to compute Euclidean distance in difference feature space. The Euclidean distance between feature vector $D1$ and $D2$ is computed as follows:

$$ED(D1, D2) = \sqrt{\sum_{b_i, b_j \in B, i<j} (d1(b_i, b_j) - d2(b_i, b_j))^2} \tag{5}$$

Bayesian Inference. In contrast to deterministic technique, probabilistic technique forms the second category. The core is the use of Bayesian inference to compute the posterior probabilities over locations. In general, an estimation is represented as a probability distribution over all the locations in the area of interest. The Bayesian inference method is used to compute a distribution conditioning on the observed signal strength. Finally, the estimated location is the one with the maximum probability in the resulting distribution.

To extend this technique with difference feature, both the representation and the Bayesian inference calculation have to be changed. All possible signal strength differences extracted from RSS are modeled as a finite observation space $O' = \{o'_1, o'_2, ..., o'_r\}$. An observation o' in the observation space O' consists of a set of signal strength differences computed from RSS. Thus, each observation o' is represented as a vector of t pairs as follows:

$$o' = \{(b_1 \times b_2, d(b_1, b_2)), ..., (b_{m-1} \times b_m, d(b_{m-1}, b_m))\} \tag{6}$$

where $b_i \times b_j$ represents the AP pair and $d(b_1, b_2)$ is the signal strength differences computed from RSS of the AP pair.

In the off-line training phase, labeled RSS data are collected at each location l_k. Signal strength difference features are extracted from these RSS and recorded at each location as observations o'. Then we build a histogram of observation for each AP pair $b_i \times b_j$ at each location l_k. This is done by constructing the conditional probability $Pr(d(b_i, b_j)|b_i \times b_j, l_k)$, which is the probability that AP pair $b_i \times b_j$ has the signal strength difference $d(b_i, b_j)$ at location l_k. By making an independence assumption among signal strength differences from different AP pairs, we multiply all these probabilities to obtain the conditional probability of receiving a particular observation o' at location l_k as follows:

$$Pr(o'|l_k) = \prod_{1 \leq i < j \leq m} Pr(d(b_i, b_j)|b_i \times b_j, l_k) \tag{7}$$

which is exactly the content of a radio map introduced before.

In the online phase, a posterior distribution over all the locations is computed using Bayesian rule:

$$Pr(l_k|o'^*) = \frac{Pr(o'^*|l_k)Pr(l_k)}{\sum_{k=1}^{n} Pr(o'^*|l_k)Pr(l_k)} \tag{8}$$

where o'^* is a new observation obtained from currently measured RSS. $Pr(l_k)$ encodes the prior knowledge about where a user may probably be. $Pr(l_k)$ can be set as the uniform distribution, assuming every position is equally likely. The estimated location l^* is the one which obtains the maximum value of the posterior probability:

$$l^* = \arg\max_{l_k} Pr(l_k|o'^*) \tag{9}$$

4 Experiment

4.1 System Setup

In order to evaluate the performance of DIFF, we establish our own wireless network environment. Our experimental test-bed is deployed on the 3rd floor of our academic building with an area of about $30m \times 17m$, covering a hallway and five rooms. We deploy 7 TENDA APs around the area to set up an IEEE 802.11b wireless network infrastructure. These APs are denoted by red stars in Figure 4 and the whole area is divided into 161 grids for signal collection, each with a size of $1m \times 1m$.

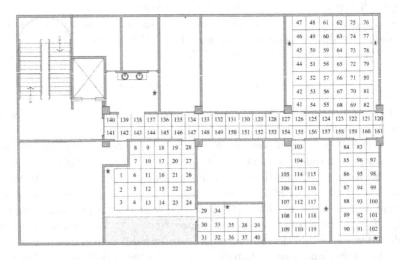

Fig. 4. The layout of the experimental test-bed in our building

We employ two types of mobile devices: an IBM R60 laptop equipped with an Intel cPro/2200GB internal wireless card and an O2 Xda Atom Life smartphone. We carry them walking around the area and stay by each grid for several seconds to collect RSS data and record the physical locations. First we carry the laptop with sample rate 5Hz and stay 40 seconds by each grid to collect a data set with 32200 samples. Then we carry the smartphone with sample rate 2Hz and stay 50 seconds by each grid to get a data set with 16100 samples. We averaged both data sets per 20 samples to reduce the effect of noise in NN algorithms.

4.2 Performance Evaluation

To test the difference feature performance over diverse devices, we use the laptop as train device and the smartphone as test device. We build the radio map based on train device's RSS values, which is used to predict test device's location with each RSS vector. The error distance is calculated by the Euclidean distance between a predicted location and its ground truth value which recorded during data collection phase. Then the evaluation results are given in term of Cumulative Probability Distribution of error distance.

Our evaluation includes the techniques of Nearest Neighbor (NN) [6] and Bayesian Inference (BI) implemented in four setups: a DIFF version (implemented as presented in Section 3), a RSS version, a RATIO version [1], and a LINEAR version extended with extra calibration effort on test device. The LINEAR version handles signal strength variance between diverse devices using linear mapping, as described in [2] [3]. The linear mapping transforms test device's RSS to match train device's RSS. The parameters for the linear mapping are found by comparing RSS collected at some locations with both devices using least squares estimation. The linear mapping is then applied to all test samples before they are forwarded to a RSS technique. Considering it's not easy to collect many measurements on test devices in practise, we use RSS collections from both devices over 10% grids in all 161 grids to calculate the linear mapping parameters. To sum up, DIFF, RSS and RATIO versions only need data of train device while LINEAR version needs extra data of test device to get the linear mapping parameters.

Figure 5 shows the experimental result of these four NN based methods. We can observe that the RSS-NN gets the worst result, indicating that variance

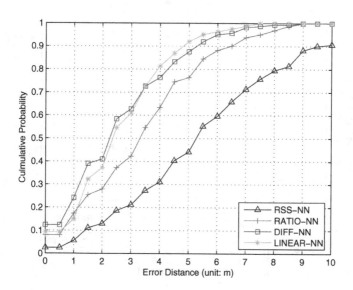

Fig. 5. Performance of NN

handling is necessary between diverse devices. Actually, the conclusions of our method and [1] is consistent with [2] [3]: it's a linear mapping between signal strength difference of diverse devices. [1] considers only the ratio term of the linear transformation function while our analysis and evaluation show the offset term is a more significant part. Therefore, as shown in Figure 5, DIFF-NN performs better than RATIO-NN and is equal to LINEAR-NN. Contrast to RATIO-NN, it improves NN accuracy with nearly 20% within an error distance of 3m. Besides, DIFF-NN can achieve equal accuracy with LINEAR-NN without calibration effort on test device, which indicates again that the offset term has an strong impact on linear transformation function.

Figure 6 shows the experimental result of these four BI based methods. We can get similar conclusion from Figure 6 as that above. However, compared with Figure 5, it can be observed that the BI accuracy is lower than NN accuracy in RSS cases. For example, within error distance of 3m, the accuracies of RSS-NN and RSS-BI are 20% and 12% separately. This differs from former works of single device [10] [15] [16], which reach the same conclusion that probabilistic techniques can reach better accuracy than deterministic techniques. The reason is that due to apparently signal strength variance, the posterior probability over locations may all be equal to zero. In this kind of situation, the first grid's location is treated as the predicted location. However, NN could always find a nearest neighbor and return its location as estimation. Therefore, NN performs better than BI for different devices. This is also consistent with conclusions in [1].

Therefore, once we have collected a complete set of RSS data from train device, the signal strength variance problem between diverse devices can be solved by using signal strength difference feature extracted from RSS without extra calibration effort on test device.

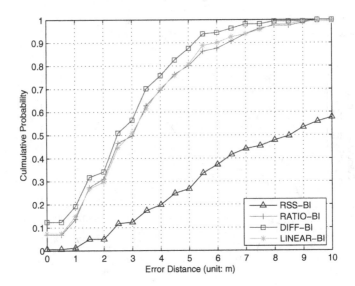

Fig. 6. Performance of BI

5   Conclusion and Future Work

In this paper, we present a calibration-free localization algorithm for handling signal strength variance between diverse devices, which we call it DIFF. The key idea is to use signal strength differences between pairs of APs instead of absolute signal strength values. We extend DIFF with two well-known localization methods: Nearest Neighbor and Bayesian Inference to evaluate the performance. Using collected data from real-world environment, we compare DIFF with other 3 methods: traditional RSS version, RATIO version and a manual LINEAR version. The results show that our method outcomes RSS and RATIO and is equal to LINEAR version.

Our work can be extended in several directions. First, we will consider to evaluate other localization techniques with DIFF and other technologies such as GSM where signal strength variance are also present. Second, we wish to reduce the computing cost of DIFF: now the dimension of difference feature is C_m^2. When the AP number m is large, selecting a subset from all APs to calculate the difference features for localization is an option. In addition, we also wish to test the validity of our proposed algorithms in a larger-scale environment.

Acknowledgements

This work is supported by the National High Technology Research and Development Program of China (863 Program, No.2007AA01Z305), the National Natural Science Foundation of China (No.60775027) and Co-building Program of Beijing Municipal Education Commission.

References

1. Kjærgaard, M.B., Munk, C.V.: Hyperbolic location fingerprinting: A calibration-free solution for handling differences in signal strength (concise contribution). In: Sixth Annual IEEE International Conference on Pervasive Computing and Communications, pp. 110–116 (2008)
2. Haeberlen, A., Flannery, E., Ladd, A.M., Rudys, A., Wallach, D.S., Kavraki, L.E.: Practical robust localization over large-scale 802.11 wireless networks. In: Proceedings of the 10th Annual International Conference on Mobile Computing and Networking, pp. 70–84 (2004)
3. Kjærgaard, M.B.: Automatic mitigation of sensor variations for signal strength based location systems. In: Hazas, M., Krumm, J., Strang, T. (eds.) LoCA 2006. LNCS, vol. 3987, pp. 30–47. Springer, Heidelberg (2006)
4. Zheng, V.W., Pan, S.J., Yang, Q., Pan, J.J.: Transferring multi-device localization models using latent multi-task learning. In: Proceedings of the Twenty-Third AAAI Conference on Artificial Intelligence, pp. 1427–1432 (2008)
5. Sun, Z., Yiqiang Chen, J.Q., Liu, J.: Adaptive localization through transfer learning in indoor wi-fi environment. In: Proceedings of the 2008 Seventh International Conference on Machine Learning and Applications, pp. 331–336 (2008)
6. Bahl, P., Padmanabhan, V.N.: Radar: An in-building rf-based user location and tracking system. In: INFOCOM, pp. 775–784 (2000)

7. Bahl, P., Padmanabhan, V.N.: Enhancements to the radar user location and tracking system. Technical report, Microsoft Research (2000)
8. Smailagic, A., Siewiorek, D.P., Joshua, A., David, K.: Location sensing and privacy in a context aware computing environment. In: IEEE Pervasive Computing, vol. 9, pp. 10–17 (2001)
9. Youssef, M., Agrawala, A.K.: The horus wlan location determination system. In: Proceedings of the 3rd International Conference on Mobile Systems, Applications, and Services, pp. 205–218 (2005)
10. Ladd, A.M., Bekris, K.E., Rudys, A., Kavraki, L.E., Wallach, D.S., Marceau, G.: Robotics-based location sensing using wireless ethernet. In: Proceedings of the Eighth Annual International Conference on Mobile Computing and Networking, pp. 227–238 (2002)
11. Youssef, M., Agrawala, A.K.: Handling samples correlation in the horus system. In: INFOCOM (2004)
12. Youssef, M., Agrawala, A.: Small-scale compensation for wlan location determination systems. In: IEEE WCNC 2003 (2003)
13. Gentile, C., Berndt, L.: Robust location using system dynamics and motion constrains. In: IEEE Conference on Communications (2004)
14. Tsui, A., Chuang, Y.H., Chu, H.H.: Unsupervised learning for solving rss hardware variance problem in wifi localization. In: Mobile Networks and Applications (2009)
15. Roos, T., Myllymäki, P., Tirri, H., Misikangas, P., Sievänen, J.: A probabilistic approach to wlan user location estimation. IJWIN 9(3), 155–164 (2002)
16. Youssef, M., Agrawala, A.: On the optimality of wlanlocation determination systems. In: Comm. Networks and Distributed Systems Modeling and Simulation Conf. (2004)

Indoor Location and Orientation Determination for Wireless Personal Area Networks

Zekeng Liang, Ioannis Barakos, and Stefan Poslad

School of Electronic Engineering and Computer Science,
Queen Mary, University of London, UK
{zekeng.liang,ioannis.barakos,stefan.poslad}@elec.qmul.ac.uk

Abstract. This paper presents a Wireless Personal Area Network (WPAN) indoor location determination system that adapts to both dynamic physical environmental conditions and human movement changes in order to find estimated user locations and their orientation. This system has been realized using the Sun SPOT sensor platform. This research identifies the challenges when deploying indoor location determination systems based upon a combination of radio signal strength indication (RSSI) and accelerometer measurements of users' mobile terminals. The experimental results show that users' indoor locations can be estimated more precisely and with greater computational efficiency compared to current systems.

Keywords: indoor location determination, user positioning, radio map, RSSI, accelerometer values, adaptive.

1 Introduction

Indoor location determination technologies have many useful pervasive computing application areas. Our main focus is towards Wireless Personal Area Networks (WPANs) applications, such as smart home service management related to users' locations, hands free local device activation, gesture based control and monitoring and assisting the elderly and disable people [1]. The most commonly used position determination method, Global Positioning System (GPS), does not work indoors because it usually requires a line-of-sight between the receiver and the transmission satellites used for positioning. Three distinguishing requirements for indoor location determination systems are the location determination accuracy (represented using the error distance between the estimated location and the actual location), the location determination precision (the repeatability of location determination) and the processing time [2], [5], [6].

The granularity for location information can vary across various applications. For instance, locating a person in a room needs more fine-grained location information whereas locating a person in a building, i.e., which room a person is in, requires more coarse-grained location information. Real-time location tracking systems require a real time response and a fixed processing time in order to locate fast moving humans or objects or to track more slowly moving objects and elderly humans. Various methods have been proposed for indoor location determination such as received signal strength indication (RSSI), time-of-arrival (TOA), time-difference-of-arrival (TDoA) and

R. Fuller and X.D. Koutsoukos (Eds.): MELT 2009, LNCS 5801, pp. 91–105, 2009.
© Springer-Verlag Berlin Heidelberg 2009

angle-of-arrival (AOA) that can be used with different types of wireless network, including WLAN, ultrasound, infrared, and Bluetooth. Techniques based upon RSSI use in WPANs are relatively simple and more robust in multipath conditions than other methods. WPAN location determination systems conventionally work in both an off-line phase to build up radio maps for a regular rectangular grid area, and an online location determination phase to receive a set of radio signal strength samples from multiple base stations positioned in different locations of a testing area to compute estimated locations based on pre-constructed radio maps.

The environment for radio-frequency (RF) based indoor location determination systems is often complex and can be affected by several static and dynamic factors. For example, static factors include the building material, the shape of the building, and items located in the building such as furniture and home and office equipment. Non-static factors represent humans and their dynamic movement in buildings as well as environmental factors such as room temperature, humidity, and pressure conditions. As a result, the calculation of the RSSI vector directly from a mobile user based on a static, previously built radio maps, will likely be affected by such static and non-static factors.

Whilst a large numbers of research applications have focused on detecting location in a rather large sized indoor environments, e.g., a size of 20 by 40 meters floor plan and with relatively coarse-grained location information results, less attention has been paid to the more challenging problem of tracking moving humans' or objects' locations within a small area that typically characterizes WPAN applications and which requires more fine-grained location information. This paper focuses on using RSSI values received from base stations located in different areas of a room to determine user location for indoor environments, while using the three axes acceleration values collected simultaneously with the RSSI values to reduce a possible location search space. The novelty of this research proposes that the location estimation accuracy and precision can be improved by the three methods based upon RSSI and three axes acceleration measurements for indoor environment. Firstly, reference points are used within the test area to adapt the current received signal strength vector to previously built radio maps in order to avoid various environmental and other conditions changes that affect the received RSSI readings. Secondly, we propose two search methods, *Maximum Search* and *Minimum Search* methods, to improve the location accuracy, location precision and in order to reduce the amount of processing time. Finally, acceleration values are used to reduce the possible search space. The system is implemented based upon multiple Sun SPOT platforms.

The rest of the paper is structured as follows. In the next section, we discuss related work. In Section 3 we present our system framework and detailed methods including data collection and results calculations. The experimental results are shown in Section 4. Finally, Section 5 discusses the results and proposes the direction of future work in order to improve the location precision and reduce the error distance.

2 Related Work

RADAR [8] is an early WLAN-based positioning system that uses a nearest neighbor algorithm and signal space method to triangulate a user's location. The system estimates the user's location by averaging the multiple nearest neighbors. The accuracy of this system is about two meters with a twenty-five percent probability, three meters with a fifty percent probability and about five meters with seventy-five percent probability. Similarly, Placelab [4] utilizes the Access Point's coordinate information in a database to predict nearby user location. The Horus [5] is an indoor WLAN location determination system that uses multiple signal strength samples from many access points allocated within a building floor to calculate a mobile device's location. The Horus system like many other WLAN location determination systems works in two phases. An offline phase builds a radio map and an online phase estimates users' locations based on the received signal strength from multiple access points and from the radio map built in advance. When searching through the radio map, the Horus system will rank the measured signal strength vector in a descending order before searching in order to find out the possibly nearest location from the particular base station. By comparing the probability the highest two estimated locations, a stopping threshold is used to determine when the system should stop searching and home in on the final estimated location.

The Adaptive Temporal Radio Maps for Indoor Location Estimation project [2] is an indoor WLAN location determination system that uses the IEEE 802.11b infrastructure. One key improvement of this system, compared to the Horus system, is that it deploys reference points for the mobile device to receive signal strength samples from the access points within the environment. This enables it to capture the dynamic relationship between signal strength values received by the reference points and the values received by the mobile device. In this approach, the system does not need to rebuild the radio map in a small place even if some of the environments have changed. Cricket [6], [7] calculates the distance between two points using TDoA from ultrasound and RF receivers. Although the accuracy can be up to the centimeter level, numerous receivers need to be deployed in the system and it is a lengthy process to instrument a physical environment to support this.

Ladd, A.M. [9] developed another indoor location determination system which is also based on IEEE 802.11b wireless Ethernet standard. The processing of the system contains the typical two working phrases, offline training and online phrase to determine the user location based on the received signal strength and on the training data set. The system achieves the accuracy of within one meter with a probability 0.64 without indicating the user orientation.

In the Smart Floor [10] system, users are identified based upon their footstep force profiles. This uses a biometric user identification system collecting information from floor tiles that are fitted with force measuring sensors. Its biometric identification system is based upon the uniqueness of each person's footstep, i.e., individual humans walk in a different way. The Geta Sandals [11] project, which is similar to Smart Floor, identifies users based upon their footsteps. RFID tags and accelerometer sensors are attached to people's sandals and this transmits their data to a central computer that extracts the footstep biometric information based upon the accelerometer values. These two systems have the similarity of identifying users' steps to understand their movements. However, they mainly focus on identifying who the users are based upon their individual uniqueness of footstep and do not focus on their movement, location and orientation.

The received signal strength can vary with respect to different receiving angles or different sending angles [12]. However, few projects consider this issue or provide user orientation information along with the location coordinate. Some projects do consider the user orientation in their radio maps, but the results are usually worse than those which do not consider user orientation in terms of location accuracy. The reason could be that those systems cannot identify the difference between some locations with different orientations that have a similar RSSI vector. The last two surveyed systems did use accelerometer sensors to detect and identity individual user footsteps. Their focus is not on detecting user movement and orientation. Furthermore, most of the projects do not consider the time needed for processing to get user location. They assume users will move slowly enough for them to finish the data processing before a user moves again.

All of the surveyed methodologies can be applied in static environments to achieve a defined level of accuracy. RSSI is affected by not only environmental variations such as temperature, humidity, the building construction material, but also by furniture, other things and by humans within it. Apart from the systems in [2] and [9], most of the existing frameworks assume that their radio maps are static and are created during an offline phase. They then assume that they will be used to estimate users' locations without any adaptation. This assumption is not practical because the values of RSSI can vary at a fixed location during different times of day and can differ from day by day using the same measure equipment - imprecision. Consequently, location estimation that relies only on a previous radio map could be imprecise. Furthermore, none of the surveyed projects use accelerometer values along with RSSI values to improve the estimated location accuracy and precision, and to provide orientation information. A solution is proposed to solve this problem by adapting the current user RSSI vector to update previous built radio maps online, as well as using both RSSI and accelerometer values for higher accuracy and precision.

3 Methodology

3.1 System Framework

The main system components and data processing flows are shown in **Fig. 1**. There are four main components in the framework; *Radio Map Manager, Radio Maps, User RSSI Converter and Location Estimator*. The *Radio Map Manager* manages the modification and access controls of radio maps and acts as a portal for other components to access the radio maps. For instance, by giving a request with parameters of one RSSI value and the base station ID, the *Radio Map Manager* can return a list of locations with orientations and their probabilities. The structure of a radio map and its data samples is shown in **Table 1**. The *spotID* stores the identity of a Sun SPOT device that receives the RSSI data from a mobile user. The *RssiReading* stores the actual RSSI value. The *timeStamp* column keeps the time to generate RSSI record and the *LocID* and *Orientation* columns record the corresponding location and direction respectively.

The processes of our framework are described as follows. Firstly, the *Radio Map Manager* works in both offline and online phases. In the offline phase, it collects the user mobile RSSI value vector along with users' actual locations and adds them into a Radio Maps database. The *Preference Point RSSI Vectors* data is stored simultaneously in the *Radio Maps*. In the online phase, depending on whether user feedback is available or not, it will update the *Radio Maps* accordingly, to improve the estimation accuracy. Secondly, the *User RSSI Vector Converter* converts the current received user

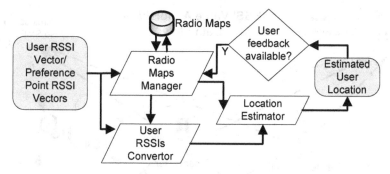

Fig. 1. System components: the arrows show the information flow in the system

RSSI vector data based on the current and existing *Preference Point RSSI Vector data* stored in database, so that the current user RSSI data can be adapted and matched to the radio maps which are built in an offline phase. *Location Estimator* searches through the radio maps to estimate the user location based upon the converted user RSSI vector. Two search methods are considered, *Maximum Search* and *Minimum Search* Methods. The details of these search methods will be given in later in this paper.

Table 1. Structure of a radio map and its data samples

spotID	Rssi Reading	timeStamp	LocID	Orientation
0014.4F01.0000.4A9E	-20	2009-03-18 12:02:26	L(2,7)	E
0014.4F01.0000.4A9E	-25	2009-03-18 12:04:36	L(1,5)	N
0014.4F01.0000.4A9E	-30	2009-03-18 12:07:49	L(3,3)	W

3.2 RSSI Data Acquisition

During the offline training phase, RSSI data is needed in order to build the radio maps in addition to its used for the online location determination. As shown in **Fig. 2**, during the offline training phase two different types of RSSI data, mobile user data and reference point RSSI data are used. *User signal broadcaster* and *Reference point broadcaster* data packets are transmitted. Wireless Base Stations that are located in different areas, e.g., corners of a room of a room, will collect the broadcast packets and extract the RSSI values along with the packet id and the time the packet was generated from the *User signal broadcaster*. This information is then sent back to the *RSSI Receiver*. User signal broadcasters will move from one location to another whereas the *Reference Point broadcasters* are static at the same locations. Only one *Reference point broadcaster* is showed in **Fig. 2**, but several of these can be installed in different places in an area to broadcast packets in order to understand the environment conditions in terms of the relationship between RSSI values and the distance between two locations.

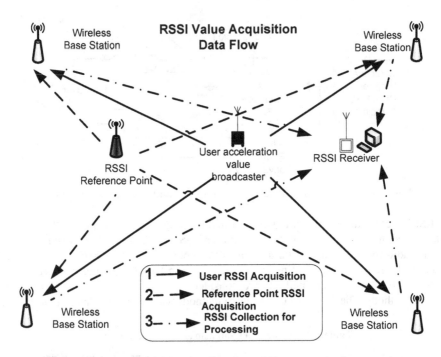

Fig. 2. RSSI vector acquisition data flow for mobile users and reference points

3.3 Adapting User RSSI Vector for Temporal Radio Maps

Environmental conditions and other changes might affect the RSSI read from different locations in a room at different times. This will lead to a variable location estimation based upon radio maps built within different time periods. There are two solutions to this problem. One is to update the whole radio map periodically to adapt to the environmental condition changes and other changes. The second solution is to adapt the user RSSI vector read during the current time period in relation to an earlier one read when the radio maps were built.

In terms of the computational processing time, updating or modifying the complete radio maps will take longer than just updating the user RSSI data vector especially when the radio map database grows larger. The computational complexity when updating radio maps is based upon regression analysis [2] and is shown in Equation (1).

$$f(n,m) = n * m * r \tag{1}$$

where n is the number of base stations to receive the mobile user RSSI values, m is the number of preference points to broadcast packets and r is the number of records in each radio map.

The computational complexity for updating the user RSSI data vector is shown in Equation (2). The computational complexity will not be affected by the number of RSSI records stored in the radio maps.

$$R(n,m) = n*m \tag{2}$$

Equations (1) and (2) show that the computational complexity of updating the radio maps can increase rapidly affected by the record number r, while the computational complexity remains the same when updating the user RSSI vector data, no matter how the history record number changes. As a result, the radio map might not be able to be updated frequently enough to reflect the environmental and other conditions changes. Hence, updates can occur daily, every several hours or every minute when significant environment changes occur. Newly observed reference RSSI values are used to compare with previous ones, in order to calculate a threshold value for updating the radio maps and the received user RSSI values.

The conversion of the user RSSI value to adapt to the temporal radio maps is shown in **Fig. 3** assuming the radio maps have been built at time $t0$ when both the reference points and user RSSI values were recorded separately. At time $t1$, both the user and reference points' RSSI values are collected. These two sets of data along with the reference points' RSSI values collected at time $t0$ will be used to calculate the equivalent user RSSI vector that would have been collected having the same environmental and other conditions at time $t0$ marked as $t1$` for use with radio maps built at $t0$. The same processes can be used to calculate the equivalent user RSSI values at a later time in order to make use of the previous built radio maps. A similar approach can be applied to update the radio maps that have been built at an earlier time.

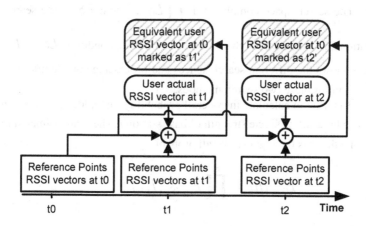

Fig. 3. Converting a user RSSI vector to adapt to temporal radio maps

3.4 Maximum and Minimum Search Methods

There are two proposed search methods used in this paper, the *Maximum Search Method* and *Minimum Search Method*. The main difference between these two methods is the search space size which will lead to a different computing complexity that affects the processing time to calculate user locations.

The *Maximum Search Method* will find out all of the corresponding locations that match a given RSSI values at different base stations. Next, it will try to match the locations from different base stations and choose the location that has the highest combined probability. So the search space complexity for the *Maximum Search Method* should be LS^n, where LS is the total locations in the room and n is the number of base stations. The computational complexity for matching all the locations from different base stations to find out the location that has the highest probability is $\prod_{i=1}^{n} LS'_i$, where LS'_i is the matched location number in base station i and n is the total number of base stations.

The *Minimum Search Method* will first sort the user RSSI vector by descending order because the higher the RSSI value, the higher chance the base station will be nearer to the mobile user, and the more likely the RSSI values match the locations stored in the pre-built radio maps. The next step is to find out all the corresponding locations that match a given RSSI value, the highest value amongst a vector, in the base station. The second base station has the second highest RSSI value that will be used to search and find out the matched locations but the search space is based upon the search results of the first base station. The same search method is used until the last base station is used. The final results will be the estimated locations with a combined probability. The search space complexity is $\prod_{i=1}^{n} LS_i$, where LS_i is the search space for base station i, n is the total number of base station, with condition $LS_0 > LS_1 > ... > LS_n$. **Fig. 4** shows the different search spaces and processing complexity to estimate the user location for these two search methods.

Equation 3 is used to calculate the overall probability that a location i exist in all R base stations where ss_i is single strength of base station i. The final estimated location will be the one that has the highest overall probability.

$$P\_all = \prod_{i=1}^{R} P(ss_i, i, l) \qquad (3)$$

For each base station, all the locations l_i are determined along with the possibilities that their RSSI values match the received RSSI value with a certain level of precision adjusted by the precision adjustment parameter ∂ :

$$| Rssi_a(l_i) - \partial | \le Rr_a \qquad (4)$$

Where $Rssi_a(l_i)$ is the RSSI value of location i in base station a, ∂ is used to control how well the returned location's RSSI value match the received RSSI value Rr_a.

$$P(l_i) = \frac{Total(l_i)}{\sum_{j \in K} Total(l_j)} \tag{5}$$

Where $P(l_i)$ is the possibility of location i, $Total(l_i)$ is the total record number of l_i, K is the returned locations' indexes aggregation.

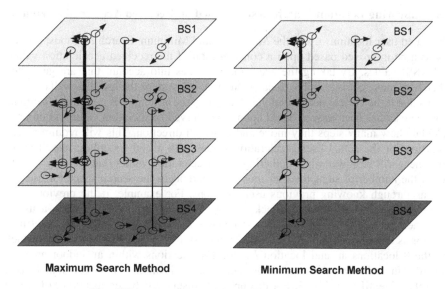

Maximum Search Method   **Minimum Search Method**

Fig. 4. Two search methods that estimate users' locations

A schematic of the use of the maximum and minimum search methods to estimate a user location amongst individual findings from base stations is shown in **Fig. 4**. Each trapezium represents a radio map for each base station. Small circles with arrows in a radio map represent possible locations with orientations that are matched to the received user RSSI data. For instance, in BS1, base station 1, the radio map could contain several locations with orientations that match a received RSSI value. Links between each map indicate the same locations with the same orientations that exist in both maps. The weight of a line represents the possibility of the computed location with orientation. If the probabilities of a location are higher among two or more adjacent maps, the line weight will be heavier. In contrast, a thinner line represents a lower possibility of the estimated location between two or more radio maps. The solid line that goes through all the radio maps represents the same location with the same orientation that exists in all the radio maps. A line that goes through all the radio maps means that the same predicted location and orientation exists in all the radio maps. Alternatively, a line that does not go through all maps represents only some of the radio maps having the same computed location and orientation among all of the stacked radio maps. Based on the received user RSSI values from the set of base stations, a set of locations that matches the RSSI values will be returned from these two search methods. The Maximum Search Method's result shown in **Fig. 4** contains two lines that go through all radio maps and six shorter lines that go through part of the radio maps. The estimated location and orientation from the Maximum Search Method is a thicker line that covers all maps. The Minimum Search Method's result shown in **Fig. 4** and includes two lines that link all maps and two other lines that connect parts of the maps. The difference between the numbers of possible results in these two methods is large, eight in the Maximum Search and four in the Minimum Search.

3.5 Improving Location Search Results Based on Received Acceleration Values

Estimated user locations using the Maximum and Minimum Search Methods, section 3.4, can be improved based upon a consideration of the received acceleration values and RSSI values. By taking the acceleration values into account that are generated when user moves, it is possible to calculate how many steps the user has walked. That means that if the current user location is known by the system, it is possible to estimate how the user has been moved based upon the observed acceleration values, in order to calculate how many steps they move and in which direction. This will require a high precision for the actual user acceleration values with a high sampling rate. The requirement will be lower when gathering user acceleration values that are only used to detect the number of steps the user moves. This reduces the search space for user locations through knowing previous user locations. For example, if the previous user location was $L(5,6)$, shown in black in **Fig. 5**, and if the system detects a user's movement, e.g., a one step walk that depends on a user's step length, the next user location is most possibly in one of the locations shown in the thick border area which are the 8 locations around location $L(5,6)$. The locations within an indoor area are modelled in such a way that the distance between two locations can be calculated based upon their identifier in **Fig. 5**. For example, the distance between location $L(3,5)$ and $L(5,6)$ can be calculated using one of the following formulas shown in Equation 6:

$$D(L(x_i, y_i), L(x_j, y_j)) = |x_i - x_j| + |y_i - y_j| \tag{6a}$$

$$D(L(x_i, y_i), L(x_j, y_j)) = \sqrt{\left(x_i - x_j\right)^2 + \left(y_i - y_j\right)^2} \tag{6b}$$

The results from both formulas will be similar when the values of parameters i and j are close, e.g., 2 and 3. The reason for this is that when a user has one or two steps, the system can detect it straightaway and estimates the user location before they will walk further away from the first location. However, the computation complexity of formula **6b** is much higher than formula **6a**, so that the first one is preferred. The experiment lab or test environment has been divided into several location points as shown in **Fig. 5** and grouped into three types. The squares in the black cross net pattern represent the locations that have been used for the radio map construction and location testing. The grey squares indicate the areas that are not accessible by users, e.g., they might be occupied by different furniture such as a meeting table. The squares in white refer to the areas are accessible by user but have not been used when building the radio map.

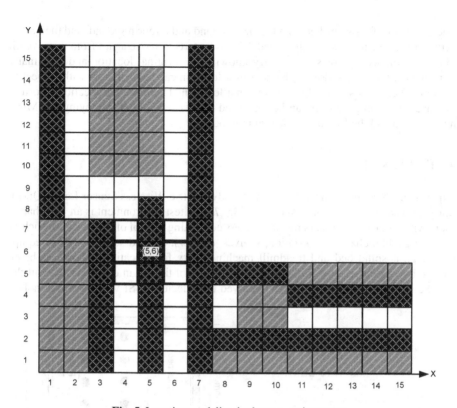

Fig. 5. Location modeling in the test environment

When users walk, related accelerometer values will change in such a way that their steps can be identified. **Fig. 6** shows how user accelerometer value changes with intermittent walking, e.g., for twelve seconds. Every step can be detected when a G-force surge of over 0.5 G appears. Before collecting the data, the accelerometer sensors need to be calibrated to have the initial value of 0G. Only one step will be counted when more than one surge above 0.5 G has been detected within a time limit of 0.5 second.

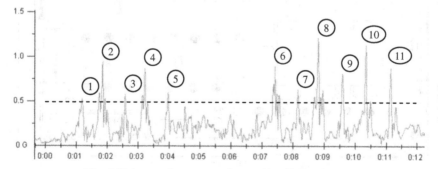

Fig. 6. Using accelerometer values to identify user's steps

The user walks four steps between the first, second and a fourth second, and then stops between the fourth and seventh second. The user starts to move again from the seventh to the twelfth second for six steps. By knowing the original location and how many steps the user has been moved, the next user location can be estimated with respect to the limited choices surrounding the original location. Therefore the location information accuracy and precision can be improved further, even after deploying the Minimum Search and the Maximum Search methods.

4   Test Results

Experiments have been carried out in the Body-Centric Wireless Sensor Lab at Queen Mary University of London as shown in **Fig. 7**. The test environment is an "L" shaped room with size of 8.66 meters by 7.89 meters containing typical office furniture such as a meeting table, chairs, workstation, shelves, drawers, etc and some body test equipment e.g., hospital bed and treadmill machine. Fifty four locations, shown as black circles, have been selected within the walking area of the lab in order to carry out the test. Seven base stations have been deployed to collected RSSI values from a mobile

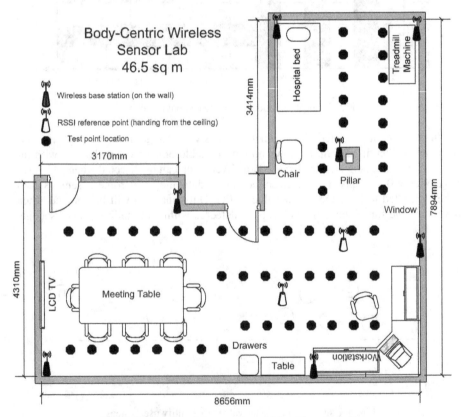

Fig. 7. Floor plan of the experiment lab with test locations

user and two RSSI reference points that broadcast packages periodically and continually during the test. The wireless base stations are installed in different corners of the room as shown in the floor plan. The two RSSI reference points are hung from the ceiling with the distance of about 30 cm in the middle of the room. This is because they need to be seen by all of the wireless base stations for reference RSSI collection. As discussed earlier, the signal strength measured from a fixed location can vary during different times of the day. These can be different even at the same time on different days. **Fig. 8** shows the received signal strength variations from a fixed location during different times of a day.

Experiments have been performed during different times in a week in order to get the average result values as the result might vary and can be affected by the static and non-static factors given in the beginning of this paper.

Table 2 shows the location precision and error distance comparison for the two search methods based on some sample locations during a user walk test. The results show that the *Maximum Search Method* provides a better location precision by 11 percent and smaller error distance by 0.13 meter but the computational processing complexity is higher. The *Minimum Search Method* gathers a lower location precision and a higher error distance compared to the first search method. The results are still promising and have a much lower computational processing complexity as described in section 3.4. For this instance, if there are 10 locations with orientations in each of seven base station radio maps that match the given RSSI value and 70 present of the locations will be matched between two radio maps, then the *Maximum Search Method*'s computational complexity will be 10^7 and less than a quarter of 10^5 for the *Minimum Search Method*. By deploying the accelerometer values to calculate the user movement steps, the error distance and location searching space can be reduced to improve the two search methods. Both search methods have an overall of about 8 percent improvement in terms of the error distance after deploying the use of the user accelerometer values. However, it will introduce some errors in the estimated user location when the two groups of estimated locations from the search using signal strength and user accelerometer values do not match at all. The later search result has to be dropped in that situation.

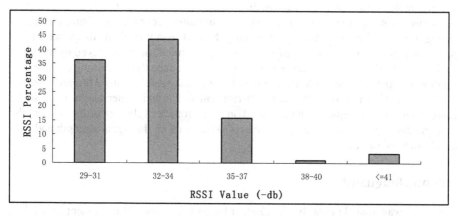

Fig. 8. RSSI values from a fixed location during different times of the day

Table 2. Location precision and error distance comparison for two search methods based on some sample locations from a user walk test

Actual Location	Maximum Search Method		Minimum Search Method	
	Estimated Location	Error Distance (m)	Estimated Location	Error Distance (m)
L(1,8,e)	L(1,8,e)	0	L(1,8,e)	0
L(3,7,s)	L(3,7,e)	0	L(3,6,w)	0.5
L(4,6,s)	L(5,8,s)	1.5	L(5,8,s)	1.5
L(5,4,w)	L(5,4,n)	0	L(5,4,n)	0
L(6,3,n)	L(5,3,n)	0.5	L(5,3,n)	0.5
L(7,2,e)	L(6,5,w)	1.5	L(6,6,e)	2.5
L(9,2,n)	L(8,2,s)	0.5	L(8,2,s)	0.5
L(11,2,w)	L(11,2,s)	0	L(11,2,s)	0
L(13,3,e)	L(13,4,w)	0.5	L(12,4,n)	0.707
	Location Precision within 0.5 m : 77.8%	Avg: 0.5625	Location Precision within 0.5 m : 66.7%	Avg: 0.6889

5   Conclusions and Future Work

Results from the two methods show that the *Maximum Search Method* searches a larger space and calculates all the possible estimated locations with orientations' possibilities to find out the most probable location and direction. The difference in computational complexity between the two search methods is significant. The *Minimum Search Method* searches a smaller space with a shorter processing time, but with a larger error distance and a lower success rate in finding the test locations. By using the user accelerometer values, both search methods have a better performance in terms of the error distance and processing time. However, they may introduce some problems when the two groups of estimated locations from the searches using the signal strength and user accelerometer values do not match at all.

Future improvement of the location determination accuracy, precision and processing time could be achieved by improving the two search methods discussed in this paper. For the *Maximum Search Method*, an improvement can be achieved by avoiding finding out all the possible locations' possibilities to reduce the processing complexity. Finding out the most suitable radio maps for the first search in the *Minimum Search Method* can also improve the search performance. Further experiments need to be carried out for user steps calculated based on accelerometer values in order to avoid the zero matching problem between two search methods using the signal strength and user accelerometer values.

Acknowledgment

This work was funded in part by the Context-based Information Management for Mobile Workers, EPSRC Industrial Case Award (Ep/C537831/1), a collaborative project between

British Telecommunications plc and Queen Mary University of London (http://www.elec.qmul.ac.uk/people/stefan/projects/Context-based-Mobile-Worker.htm).

References

1. Poslad, S.: Ubiquitous Computing: Smart Devices, Environments and Interaction, pp. 350–365. Wiley, Chichester (2009)
2. Yin, J., Yang, Q., Ni, L.: Adaptive Temporal Radio Maps for Indoor Location Estimation. In: Third IEEE International Conference on Pervasive Computing and Communications, PerCom 2005, pp. 85–94 (2005)
3. Ni, L.M., Liu, Y., Lau, Y.C., Patil, A.P.: LANDMARC: Indoor location sensing using active RFID. In: Proceedings of IEEE PerCom 2003, Dallas, TX, USA (2003)
4. LaMarca, A., Chawathe, Y., Consolvo, S., Hightower, J., Smith, I.E., Scott, J., Sohn, T., Howard, J., Hughes, J., Potter, F., Tabert, J., Powledge, P., Borriello, G., Schilit, B.N.: Place Lab: Device Positioning Using Radio Beacons in the Wild. In: Gellersen, H.-W., Want, R., Schmidt, A. (eds.) PERVASIVE 2005. LNCS, vol. 3468, pp. 116–133. Springer, Heidelberg (2005)
5. Youssef, M.A., Agrawala, A., Udaya Shankar, A.: WLAN location determination via clustering and probability distributions. In: Pervasive Computing and Communications 2003 (PerCom 2003), pp. 143–150 (2003)
6. Priyantha, N.B., Chakraborty, A., Balakrishnan, H.: The cricket ocation-support system. In: MobiCom 2000: Proceedings of the 6th annual international conference on Mobile computing and networking, pp. 32–43. ACM Press, New York (2000)
7. Wang, Y., Goddard, S., Perez, L.C.: A study on the cricket location-support system communication protocols. In: IEEE International Conference on Electro/Information Technology, 2007, pp. 257–262 (2007)
8. Bahl, P., Padmanabhan, V.N.: RADAR: An In-Building RF-based User Location and Tracking System. In: Proc. IEEE Nineteenth Annual Joint Conference of the IEEE Computer and Communications Societies (INFOCOM 2000), Tel Aviv, Israel, pp. 775–784 (2000)
9. Ladd, A.M., Bekris, K.E., Rudys, A.P., Wallach, D.S., Kavraki, L.E.: On the feasibility of using wireless ethernet for indoor localization. IEEE Trans. Rob. and Auto. 20(3), 555–559 (2004)
10. Orr, R.J., Abowd, G.D.: The smart floor: a mechanism for natural user identification and tracking. In: CHI 2000 extended abstracts on Human factors in computing systems, The Hague, The Netherlands, pp. 275–276. ACM, New York (2000)
11. Yeh, S., Chang, K., Wu, C., Chu, H., Hsu, Y.: GETA sandals: a footstep location tracking system. Personal Ubiquitous Comput. 11, 451–463 (2007)
12. Krishnakumar, A.S., Krishnan, P.: The theory and practice of signal strength-based location estimation. In: International Conference on Collaborative Computing: Networking, Applications and Worksharing, p. 10 (2005)

Localize Vehicles Using Wireless Traffic Sensors

Peng Zhuang and Yi Shang

University of Missouri, Columbia MO, USA

Abstract. Recently, wireless traffic sensors present themselves as a low cost and non-intrusive alternative to wired traffic sensors. We propose a vehicle localization method that utilizes the signals of the wireless sensors. A vehicle is equipped with a receiver and overhears the geo-tagged packets transmitted by wireless traffic sensors. An onboard computer then computes the distribution of possible vehicle locations using an algorithm based on the principles of particle filtering. In our simulation, the proposed method outperforms the proximity centroid method by an average of 79%.

1 Introduction

Vehicle localization and tracking have many applications, including vehicle navigation, theft detection, roadside assistance, etc. Most vehicle localization systems are based on GPS with map matching, which is accurate only up to about 10 meters. This is particularly a problem when the vehicle is at dense road networks, such as highway junctions. GPS units also suffer from the problem of "lose of signals" such as when the vehicles are in tunnels or under trees. Dead reckoning is commonly used when GPS signals are lost where a vehicle locates itself based on its last known GPS location and the speed it travels afterwards. Dead reckoning has even higher location errors and the errors are cumulative.

In most US cities, traffic sensors are used to monitor traffic, measure pavement wearing, and provide information for adaptive traffic signal control systems. Recently, wireless sensors have emerged as a low cost and non-intrusive alternative to wired sensors [5] [6]. Comparing to traditional sensors such as the embedded loops, they are small in size, operate on batteries, and can last up to 10 years. The wireless signal transmitted by the sensor devices provides a new way of vehicle localization.

In this paper, we present a system that achieves vehicle localization and tracking by combining the geo-tagged sensor information and vehicle's own speedometer readings. The algorithm is based on the principle of particle filtering. It is shown to achieve an accuracy of 1 2 meters. When compared to the algorithm of proximity centroid, its achieves 79% higher accuracy.

The rest of the paper is organized as follows. In section 2, we survey related works. In section 3, we present the system design and the algorithm. In section 4, we discuss the simulation results and in section 5 we reach our final remarks.

R. Fuller and X.D. Koutsoukos (Eds.): MELT 2009, LNCS 5801, pp. 106–114, 2009.
© Springer-Verlag Berlin Heidelberg 2009

2   Related Works

The problem of tracking mobile targets using sensor networks has been extensively studied in the literature. Our proposed method is inspired by the study of Monte Carlo based localization [1] [3] [4]. Hu et. al. have proposed a method called MCL that improves localization accuracy by utilizing the mobility information (maximum speed) [4]. Rudafshani et. al. have proposed two methods called MSL and MSL* that outperform MCL [3]. The improvements include faster convergence and more robust against the decrease of location beacon (seed) density. The improvements of MSL and MSL* are mainly contributed by the use of neighbors' localization estimates. Klingbeil et. al. has proposed a method that utilizes more detailed mobility information, including speed, heading, and map to achieve higher accuracy [1]. Another class of methods is based on the Kalman filter. A Kalman filter uses the exactly same process as the sequential Monte Carlo filter but provides analytical results instead of simulation results. To use a Kalman filter to localize a mobile target, the mobility model has to be confined to a linear model. It is not suitable when the target changes its speed and/or direction. Kusy et. al. have proposed a least-squares optimization based improvements to traditional Kalman filter that achieves almost 50% higher accuracy [2].

3   System Design

3.1   System Setup

The SenSys vehicle detection sensor mote is used as an example for this study [7]. The sensors are embedded under pavements. They can detect vehicle passing by and communicate only to above ground receivers. Two sensors are needed per lane for vehicle speed detection. The communication range of a sensor varies according to the height of the receiver. An access point is mounted on a roadside pole that receives packets from embedded sensors. When the height of an access point is 2.4 meters, the maximal communication range is about 23 meters. For simplicity, we assume an access point can cover a road segment of length D (e.g., 46 meters in a typical setup). An embedded sensor can talk to passing vehicles with a radius r. Since the receivers on the vehicle will be on a lower height than the access point, r is assumed to be smaller than $D/2$. We assume a vehicle knows r or the distribution of r. Such information can either be pre-installed or be included in each packet broadcasted by a sensor.

The time is divided in discrete time steps. Each sensor knows its exact geographic coordinates and periodically broadcast a location beacon packet containing the coordinates. The coordinates can be loaded when they are installed. Each vehicle has a receiver and is constantly listening to the location beacon packet. For simplicity, each vehicle's coordinates are represented by the coordinates of its receiver. We assume all receivers can only be in the center line of any lanes except when the vehicle is switching lanes. Note that our method can be easily adopted in situations with different receiver placements. A vehicle

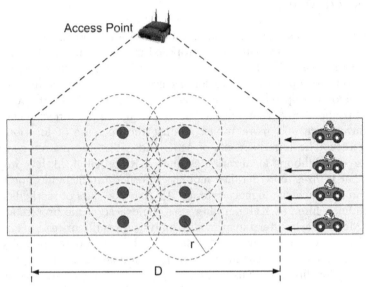

Fig. 1. Vehicle localization system setup

can hear a location beacon packet only if its receiver's distance to the sensor is less than or equal to r (the packet may still be missed according to certain probability distribution).

It is critical to reduce the communication cost of the sensor device. It is achieved by the following means:

- Since the vehicles' batteries have more power and are charged when the vehicles are moving, vehicles' receivers are set to be constantly listening and the sensors only periodically broadcast short location beacons.
- The location information is embedded in ordinary traffic reporting data packets, which reduces the overhead of the localization system.

3.2   Available Information

The in-vehicle localization system may use the following information:

1. **Location beacon:** the vehicle timestamps each received location beacon packet. A record of location beacon is given by

$$(time\ step,\ x\ coordinate,\ y\ coordinate). \tag{1}$$

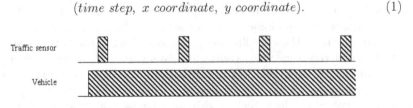

Fig. 2. Communication between a vehicle and a wireless sensor node

2. **Map information:** the vehicle knows the accurate road map, including the number of lanes and the width of each lane.
3. **Vehicle speed:** the in-vehicle localization system acquires the speed information from the vehicle's speedometer. A record of speed is given by

$$(time \ step, \ distance \ moved \ in \ this \ time \ step). \tag{2}$$

Note if a vehicle switches lanes in this time step (we assume all lane switching can be finished within one time step), a speed record contains the distance moved due to lane switching. US regulations require the error of a speedometer is at most 5%. Depending on various factors (mostly the tire diameter), the error can actually be larger. We assume the error of the speedometer is drawn from a zero mean Gaussian distribution. Since the tire diameter is mostly unchanged during a localization process, we assume the error is persistent.
4. **Vehicle heading:** the localization system acquires the moving direction of the vehicle.
5. **Radio signal strength (RSS):** RSS is not reliable in urban traffic environment due to the absorbing, blocking, and reflecting effects caused by the surrounding vehicle bodies. However, to some extent RSS can be combined with other information to improve the localization accuracy.
6. **GPS signal:** the in-vehicle localization system may carry a GPS receiver. In open space the GPS coordinate is accurate to about 10 meters. When combined with other information, the accuracy can be improved.

In this paper, we discuss a localization algorithm that utilized information 1 3 and leave the rest for future works.

3.3   Vehicle Localization Algorithm

Vehicle state :   A vehicle state S_t at time step t consists of

$$(x_t, y_t, \varepsilon(v), \theta), \tag{3}$$

where
   x_t, y_t: vehicle's coordinates at time t,
   $\varepsilon(v)$: a persistent speedometer error,
   θ : a vehicle's heading along the road.
We also add a random noise $\varepsilon_t(v)$ as the vehicle's x coordinate estimation error at time t since sometimes the filtering condition in the correction step leaves no valid samples

Initialization: The system initializes when a vehicle hears the first location beacon z_0. It estimates its location distribution $p((x_0, y_0)|z_0)$, which can simply be a circle of radius r surrounding the corresponding sensor. It can also be refined using the map information.
   In the sequential Monte Carlo filter, N equally weighted samples $(S_0^{[i]}, w_0^{[i]})$ are created as:

$x_0^{[i]}, y_0^{[i]}$: drawn from $p((x_0, y_0)|z_0)$

$\varepsilon_0^{[i]}(v)$ and $\varepsilon_0^{[i]}(x)$: drawn from two zero mean Gaussian distributions

$\theta^{[i]}$: randomly chosen from 1 or -1.

The weight of all samples sums up to unity. We have

$$\sum_{i=1}^{N} w_0^{[i]} = 1. \tag{4}$$

Localization: After the time step of initialization, in each time step the localization method is carried out. Each localization step k contains the following two sub-steps:

1. **Prediction:** Every previous sample $(S_{k-1}^{[i]}, w_{k-1}^{[i]})$ is replaced by a new sample $(\tilde{S}_k^{[i]}, \tilde{w}_k^{[i]})$ according to one of the following two mobility process models:

   A. A model that assumes the vehicle does not switch lanes

   $$\binom{x}{y}_k^{[i]} = \binom{x}{y}_{k-1}^{[i]} + \theta^{[i]} \bullet [\tilde{v}_k^{[i]} - \varepsilon^{[i]}(v)] \bullet \binom{1}{0} - \varepsilon_k^{[i]}(x) \bullet \binom{1}{0}, \tag{5}$$

   where
   $\varepsilon_k^{[i]}(x)$ is drawn from a zero mean Gaussian distribution,
   $\tilde{v}_k^{[i]}$ is the measured speed from the speedometer.

   B. A model that assumes the vehicle switches lanes

   $$\binom{x}{y}_k^{[i]} = \binom{x}{y}_{k-1}^{[i]} + \binom{\theta^{[i]}}{\tilde{\phi}_k^{[i]}} \bullet \binom{\tilde{v}_k^{[i]}(x)}{\tilde{v}_k^{[i]}(y)} - \varepsilon_k^{[i]}(x) \bullet \binom{1}{0}, \tag{6}$$

   where
   $\tilde{\phi}_k^{[i]}$ is the lane switching direction randomly chosen between 1 or -1,
   $\tilde{v}_k^{[i]}(x), \tilde{v}_k^{[i]}(y)$ are the vehicle speed along the road and perpendicular to the road, respectively. They satisfy

   $$[\tilde{v}_k^{[i]}(x)]^2 + [\tilde{v}_k^{[i]}(y)]^2 = [\tilde{v}_k^{[i]} - \varepsilon^{[i]}(v)]^2. \tag{7}$$

   For simplicity, we assume $\tilde{v}_k^{[i]}(y)$ is the width of one drive lane.

   During the prediction step, the method randomly chooses between the two models by an arbitrary probability.

2. **Correction:** The weights are updated as

   $$w_k^{[i]} = \tilde{w}_k^{[i]} \bullet p(z_k|\tilde{x}_k^{[i]}), \quad \sum_i w_k^{[i]} = 1. \tag{8}$$

   z_k is the set of observations. We have

   $$p(z_k|\tilde{S}_k^{[i]}) \propto h(\tilde{S}_k^{[i]}), \tag{9}$$

   where

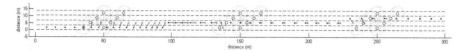

Fig. 3. 3 group of traffic sensors (represented by "o" with a circle denoting the communication range) are deployed in a road segment of 300 meters. The dots are the vehicle's trajectory, the "x" marks are the tracking results. A line between a dot and a "x" shows the localization error.

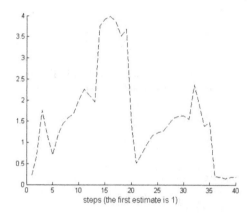

Fig. 4. The localization error of the test case shown in figure 3, shows estimation error along the road

$$h(\tilde{S}_k^{[i]}) = \prod_m f(P_m - (\tfrac{\tilde{x}}{\tilde{y}})_k^{[i]}) \prod_n g(P_n - (\tfrac{\tilde{x}}{\tilde{y}})_k^{[i]}) \qquad (10)$$

where
m is the sensors that the vehicle has heard in step k,
n is the sensors that the vehicle has not heard in step k, but has heard in step $k - 1$ or before,
P_m is the position of sensor m. We also have
$$f(d) = \begin{cases} 1, & d \le r, \\ 0, & d > r, \end{cases} \text{ and}$$
$$g(d) = \begin{cases} 0, & d \le r, \\ 1, & d > r. \end{cases}$$

3. **Re-sampling:** This step removes samples with lower weights. A new set of N samples are drawn from all samples $\tilde{S}_k^{[i]}$ with a probability $w_k^{[i]}$.

4   Simulation Results

The simulation is conducted in MATLAB with setups similar to figure 3. The communication range from sensor to vehicle is 5 meters (since the sensor is

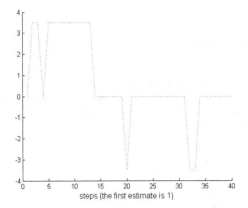

Fig. 5. The localization error of the test case shown in figure 3, shows estimation error at the direction perpendicular to the road direction

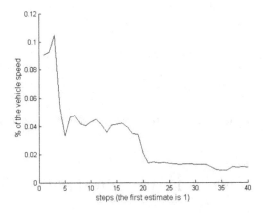

Fig. 6. The estimation error on the speedometer bias of the test case shown in figure 3

embedded and low power). The vehicle has 10% probability of switching lanes. Vehicle's initial speed is 30mph. Vehicle's acceleration is randomly drawn from a zero mean Gaussian distribution whose standard deviation is 10% of the initial speed. Speedometer's error is also drawn from a zero mean Gaussian distribution whose standard deviation is 10% of the initial speed. Sensor's packet broadcast frequency is 2Hz. Particle filtering uses 400 particles. $\varepsilon(v)$ is drawn from the same distribution as the speedometer error distribution. $\varepsilon_t(x)$ is drawn from a zero mean Gaussian distribution whose standard deviation is 0.5 (meter).

Figure 4 shows localization error along the road. When the vehicle is close to a sensor, the localization error drop to below 2 meters. In comparison, the impact of sensors on the direction perpendicular to the road (estimate which lane the vehicle is traveling in) is less significant (shown in figure 5). Figure 6 shows the estimation error on the speedometer bias. The estimates converge after 20 steps and are reduced from 10% to about 1%.

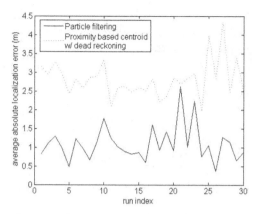

Fig. 7. From 30 runs from random maps, compared with a simple proximity beacon method

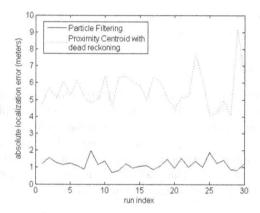

Fig. 8. Localization error when the vehicle knows the exact communication range

Next, we compare the proposed method to the proximity centroid method. In the proximity centroid, a simple dead reckoning algorithm is performed when the vehicles are outside the range of any sensors. Figure 7 illustrate the improvements achieved by the proposed methods. In all 30 test cases, the proposed method outperforms the proximity centroid method with an average improvement of 61% (1.1 meters to 2.8 meters).

In the last set of experiments, we introduce a random error to the communication radius estimation. It is reasonable since sensors' exact communication ranges are often influenced by unpredictable environmental factors. The communication range of each sensor is drawn from a Gaussian distribution $N(10, 2)$. Figure 4 8 shows the results where the vehicle knows the exact communication ranges. In the experiment corresponding to figure 9, the vehicle only knows the distribution of the communication ranges and the communication ranges are treated as a variable and are sampled in the Monte Carlo simulation. Although

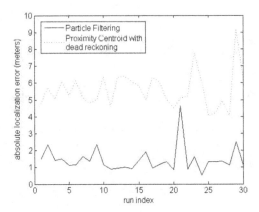

Fig. 9. Localization error when the vehicle knows only the communication range distribution

the proposed method yields more errors when more uncertainties are introduced, it still outperforms the proximity centroid method by an average of 73%.

5 Conclusion

In this paper, we propose a vehicle localization method based on Monte Carlo simulation. The method utilizes signals from wireless traffic sensors and has minimal communication overhead. In our simulation it achieves higher accuracy than both the GPS and the proximity centroid. The deployment of the wireless sensors, and the uncertainty of the communication range may limit its usage. We recommend to incorporate this method as an additional means to GPS based localization to provide higher accuracy when possible.

References

1. Klingbeil, L., Wark, T.: A wireless sensor network for real-time indoor localization and motion monitoring. In: Proc. International Conference on Information Processing in Sensor Networks (IPSN), pp. 39–50 (2008)
2. Kusy, B., Ledeczi, A., Koutsoukos, X.: Tracking mobile nodes using RF Doppler shifts. In: Proc. Conference On Embedded Networked Sensor Systems (SenSys), pp. 29–42 (2007)
3. Rudafshani, M., Datta, S.: Localization in wireless sensor networks. In: Proc. International Conference on Information Processing in Sensor Networks (IPSN), pp. 51–60 (2007)
4. Hu, L., Evans, D.: Localisation for mobile sensor networks. In: Proc. International Conference on Mobile Computing and Networking (MobiCom), pp. 45–57 (2004)
5. http://www.sensysnetworks.com/
6. Cheung, S.-Y., Coleri Ergen, S., Varaiya, P.: Traffic surveillance with wireless magnetic sensors. In: Proc. 12th Intelligent Transportation System World Congress, San Francisco, California (November 2005)
7. Sensys System Overview, http://www.sensysnetworks.com/system.html

On the Feasibility of Determining Angular Separation in Mobile Wireless Sensor Networks

Isaac Amundson, Manish Kushwaha, and Xenofon D. Koutsoukos

Institute for Software Integrated Systems (ISIS)
Department of Electrical Engineering and Computer Science
Vanderbilt University
Nashville, TN 37235, USA
isaac.amundson@vanderbilt.edu

Abstract. Mobile sensors require periodic position measurements for navigation around the sensing region. Such information is often obtained using GPS or onboard sensors such as optical encoders. However, GPS is not reliable in all environments, and odometry accrues error over time. Although several localization techniques exist for wireless sensor networks, they are typically time consuming, resource intensive, and/or require expensive hardware, all of which are undesirable for lightweight mobile nodes. We propose a technique for obtaining angle-of-arrival information that uses the wheel encoder data from the mobile sensor, and the RF Doppler-shift observed by stationary nodes. These sensor data are used to determine the angular separation between stationary beacons, which can be used for navigation. Our experimental results demonstrate that using this technique, a robot is able to determine angular separation between four pairs of sensors in a 40 x 40 meter sensing region with an average error of 0.28 radian.

1 Introduction

Until recently, mobile wireless sensors had little control over their own movement, and were typically mounted on mobile objects for purposes of identification, tracking, and monitoring. This is now no longer the case; with the emergence of small-footprint wireless sensors such as [1] and [2], nodes are able to traverse the sensing region under their own control. This has numerous advantages, such as enabling targeted coverage [3] and connecting disjoint sensor networks [4].

Arguably one of the biggest challenges for mobile sensors is navigation, where the mobile node must reach point B from point A. For the most basic wheeled mobile robots (WMRs), navigation is typically solved using odometry, whereby the robot monitors the angular velocity of each wheel to approximate the distance traveled over a given time period. The angular velocity is often determined from feedback from optical encoders mounted on each wheel. The advantage of optical encoders is that they are small and can be mounted on almost any type of WMR. Most other sensors used for navigation (e.g., GPS, laser rangefinders, sonar, etc.) are either too large, heavy, expensive, or require too much power to

R. Fuller and X.D. Koutsoukos (Eds.): MELT 2009, LNCS 5801, pp. 115–127, 2009.
© Springer-Verlag Berlin Heidelberg 2009

operate over extended periods of time. When operating on a clean, level surface, optical encoders can be quite accurate. However, most environments contain dust that can interfere with the encoder readings. Additionally, odometry rapidly accrues error on uneven terrain due to wheel slippage and low tire pressure.

Recent work [5] has shown that navigation is possible without knowing the current position of the mobile sensor. In fact, all points in a plane are reachable if the angular separation between two pairs of beacons can be determined. In many situations, navigating without having to determine position is advantageous because most localization methods require extensive PC processing, have high localization latency, and require the positions of infrastructure nodes to be known. In previous work [6], we used the Doppler shift in radio transmission frequency as control feedback to drive a robot. Although the robot was able to accurately navigate the sensing region, the localization algorithm relied on the use of an extended Kalman filter (EKF) for noisy frequency measurements. The EKF was run in realtime on the robot, but its large size required execution on a separate mote, which communicated with the robot controller over a wireless interface.

In this paper, we present a method for determining angular separation that only requires the sensor radio and wheel encoders, both of which are common to robotic wireless sensors, and hence no additional hardware is required. Our method uses the Doppler shift in radio frequency and the instantaneous velocity of a mobile node transmitting a pure sinusoidal signal to derive the angular separation between infrastructure nodes surrounding the sensing region. Our method does not require the positions of the infrastructure nodes, or the initial position of the mobile node, to be known. Because this method is intended for use with resource-constrained mobile sensors, it is rapid and "mote-able" (i.e., the algorithm runs entirely on the mote; no offline or PC-based processing is involved). We show using real-world experimental results and in simulation that this method is accurate with an average angular separation error of 0.28 radian.

The remainder of this paper is organized as follows. In Section 2, we describe our problem statement, followed by our method for angular separation estimation in Section 3. Our implementation on a mobile wireless sensor platform is described in Section 4. Experimental results are then presented in Section 5. Section 6 concludes.

2   Problem Formulation

Consider a sensing region that contains multiple infrastructure nodes, as well as a mobile sensor that needs to travel from point A to point B. This scenario is illustrated in Figure 1.

In order to navigate toward point B, we need to know which direction to drive in, and for that we need to have some idea of where we are. Localization in wireless sensor networks is a well-studied problem, and several different approaches have been developed [7], [8]. Typically, localization is accomplished by using a set of coordinated nodes, whereby one or more nodes emit a signal, such

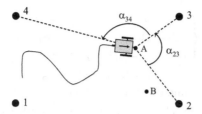

Fig. 1. A mobile sensor moves through the sensing region. The mobile node navigates based on angular separation of beacons (numbered 1 through 4).

as radio, ultrasound, infrared, or acoustic. Node position is then computed based on properties of signal arrival, such as time-of-arrival, time-difference-of arrival, received signal strength, angle-of-arrival, or by other methods.

However, localization is not necessary for navigation. In [5], a navigation method is presented that enables mobile entities to reach any position on a plane based on the angular separation between two pairs of beacons. Figure 2 illustrates the setup, simplified for clarity. In the actual setup, at least three beacons are required for navigation. In the figure, the current position of the mobile node is denoted by C, and the destination position by D. We do not know these actual positions, nor do we know the positions of nodes R_i and R_j. The only information we have is the angle $\angle R_i C R_j$ at our present position, denoted by α_{ij}, as well as the angle $\angle R_i D R_j$ at our destination, denoted by α'_{ij}. What we require is a control law that takes these angles as input and provides us with the necessary motion vector to reach our destination.

With current and destination angle information, the objective is to minimize the difference, $\Delta\alpha_{ij} = \alpha'_{ij} - \alpha_{ij}$, between the two. When $\Delta\alpha_{ij} = 0$, the mobile node will have reached its destination. For example, if $\Delta\alpha_{ij} > 0$, this means that $\angle R_i D R_j > \angle R_i C R_j$. Therefore, the mobile node should move along a vector that increases α_{ij}. The unit length bisector vector δ_{ij} is one such vector. Based on this reasoning, the complete motion vector developed in [5] takes the form

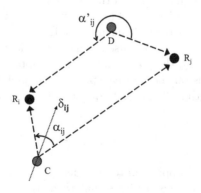

Fig. 2. Navigation without localization based on beacon angular separation

$$\mathbf{M_{ij}} = \begin{cases} \Delta\alpha_{ij} \cdot \delta_{\mathbf{ij}}, & -\pi \leq \Delta\alpha_{ij} \leq \pi \\ (2\pi - \Delta\alpha_{ij}) \cdot \delta_{\mathbf{ij}}, & \Delta\alpha_{ij} > \pi \\ (-2\pi - \Delta\alpha_{ij}) \cdot \delta_{\mathbf{ij}}, & \Delta\alpha_{ij} < -\pi \end{cases}$$

As the mobile node moves along the bisector $\delta_{\mathbf{ij}}$, it is actually following a segment of a hyperbolic curve, defined by the two foci R_i and R_j. By following such a curve, the mobile node is guaranteed to reach the arc R_iDR_j. With only two beacons, position D cannot be identified on this arc because for every point, $\mathbf{M}$ becomes 0. However, with three beacons, D is constrained to lie on two or more arcs. In other words, $\mathbf{M} = \mathbf{M_{ij}} + \mathbf{M_{jk}} + \mathbf{M_{ki}}$, and the destination D is reached when $\forall ij : \Delta\alpha_{ij} = 0$.

Often, angle information is determined using cameras, microphone arrays, or light pulses, all of which are not ideal for lightweight mobile sensors. We would like to estimate angular separation using only hardware that is widely available on sensor nodes. Specifically, we would like to be able to obtain this angle information using the sensor node radio and the optical encoders on the wheels. Although we have not implemented the above navigation method for our current work, we refer to it here as motivation for our angular separation estimation technique, and have plans to integrate it in the near future.

3   Estimation of Angular Separation

In this section, we describe the design of our angular separation estimation method for mobile sensors.

A mobile node, T, moving through the sensing region with velocity $\mathbf{v}$ and heading φ, collects angular velocity data from its wheel encoders. For WMRs with 2-wheel differential steering, the relationship between the robot speed and the wheel angular velocities is

$$|v| = \frac{r(\omega_r + \omega_l)}{2} \tag{1}$$

where the speed $|v|$ is the magnitude of the velocity $\mathbf{v}$, r is the wheel radius, and ω_r and ω_l are the right and left wheel angular velocities, respectively.

As the mobile node moves, it transmits an RF sinusoidal signal, which is observed by the receiver nodes. Because the mobile node is moving with respect to the stationary receivers, the RF signal will be Doppler-shifted. The amount of Doppler shift depends on the relative speed of the mobile and anchor nodes, as well as the wavelength and carrier frequency of the signal. This relationship is formalized as

$$f_i = f_{carrier} - \frac{v_i}{\lambda} \tag{2}$$

where f_i is the observed Doppler-shifted frequency at receiver R_i, $f_{carrier}$ is the transmission frequency of the carrier signal with wavelength λ, and v_i is the relative speed of mobile node T with respect to receiver R_i.

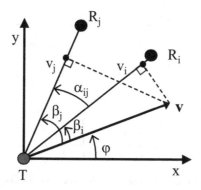

Fig. 3. Geometry of simplified setup for determining angular separation between two receivers

Figure 3 illustrates the geometry of a simplified setup. For now we will only consider two receiver nodes, R_i and R_j. The problem is to estimate the angular separation α_{ij} between the two receiver nodes based on the measured quantities of ω_r, ω_l, f_i, and f_j, and known values $f_{carrier}$ and λ.

The relative speed, v_i, between the mobile node and receiver R_i is the scalar value resulting from the projection of $\mathbf{v}$ onto the position vector $\overrightarrow{TR_i}$, as

$$v_i = |v|cos\beta_i \tag{3}$$

where the speed of the mobile node, $|v|$, has a negative sign if T is moving toward R_i and positive otherwise, and β_i is the angle between the velocity vector $\mathbf{v}$ and the position vector $\overrightarrow{TR_i}$.

The relative speed is related to the received Doppler-shifted signal. By rearranging Equation (2), we have

$$v_i = \lambda(f_{carrier} - f_i) \tag{4}$$

Combining Equations (3) and (4), and rearranging, we can calculate β_i by

$$\beta_i = cos^{-1}\left(\frac{\lambda(f_{carrier} - f_i)}{|v|}\right) \tag{5}$$

Angular separation between two receiver nodes R_i and R_j can then be computed by subtracting one bearing from the other, as

$$\alpha_{ij} = \beta_j - \beta_i \tag{6}$$

The error in computing the bearing β will vary due to the nonlinearity of Equation (5). Figure 4a shows the structure of the inverse cosine function ($y = cos^{-1}(x)$), and its derivative is pictured in Figure 4b. We can see that at the limits ($-0.8 \geq x \geq 0.8$), a small error in x will result in a large error in $cos^{-1}(x)$. To avoid this problem, we examine the argument to the inverse cosine, and if

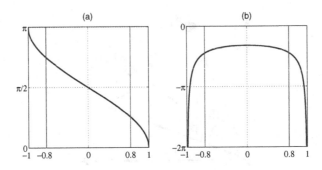

Fig. 4. (a) The inverse cosine function and (b) its derivative

too large or small, discard the sensor data for the current measurement round. In practice, we found this gives us a marginal error reduction of approximately 2°, or 11%.

3.1   Frequency Estimation Using Resource-Constrained Hardware

Typical low-cost sensor hardware supports radios that transmit in the 400 MHz — 2.6 GHz range. These radios have a received signal strength indicator (RSSI) pin which can be accessed from software, however, we cannot sample the RSSI fast enough to determine the carrier frequency, $f_{carrier}$, of the signal. Instead, we use radio interferometry [9], in which a second node transmits a signal at a slightly lower frequency such that the two transmitted signals interfere, creating a low-frequency beat signal, f_{beat}. The assistant transmitter can be positioned anywhere in or near the sensing region, as long as it is stationary, and its signal can reach all receiver nodes. The beat signal, which can be as low as a few hundred Hertz (350 Hz in our case), can be sampled by making successive reads of the RSSI.

Another issue with the inexpensive radio chip is that the transmission frequency can vary from the desired frequency by up to 65 Hz. For this reason, we treat the transmission frequency as a random variable, which results in the beat frequency, f_{beat}, being a random variable as well. This poses a challenge, because we require knowledge of the beat frequency to compute the receiver bearings. In order to solve Equation (5), we use maximum likelihood (ML) estimation [10].

For ML estimation, we rewrite Equation (5) as

$$f_i = F(\beta_i, f_{beat}) + \epsilon_i$$

$$= f_{beat} - \frac{|v|}{\lambda} \cos \beta_i + \epsilon_i$$

where $\epsilon_i \sim \mathcal{N}(0, \sigma_f)$ is the Gaussian noise in the observed Doppler-shifted frequency. The negative log-likelihood for f_i is given by

$$\ell_i(f_{beat}, \beta_i) = -\ln p(f_i | f_{beat}, \beta_i) = \frac{\| f_i - F(\beta_i, f_{beat}) \|^2}{\sigma_f^2}$$

Assuming N receivers, the combined negative log-likelihood for $f_i, i = 1, \cdots, N$ is given by

$$\ell(f_{beat}, \beta_1, \cdots, \beta_N) = -\ln p(f_1, \cdots, f_N | f_{beat}, \beta_1, \cdots, \beta_N)$$

$$= -\ln \prod_{i=1}^{N} p(f_i | f_{beat}, \beta_i)$$

$$= \sum_{i=1}^{N} \ell_i(f_{beat}, \beta_i)$$

$$= \sum_{i=1}^{N} \frac{\| f_i - F(\beta_i, f_{beat}) \|^2}{\sigma_f^2}$$

The ML estimate can be obtained by minimizing the negative log-likelihood using the following

$$\frac{\partial \ell(f_{beat}, \beta_1, \cdots, \beta_N)}{\partial f_{beat}} = 0$$

The partial derivative leads to the following result for the ML estimate for the beat frequency

$$\widehat{f}_{beat} = \frac{1}{N} \sum_{i=1}^{N} f_i + \frac{|v|}{\lambda N} \sum_{i=1}^{N} \cos \beta_i$$

Note that the ML estimate, $\widehat{f}_{beat}$ is in terms of $\beta_i, i = 1, \cdots, N$. To solve for the angles, we iteratively compute the ML estimate and the angles. The two iterative steps are given below.

1. Computing the angles:

$$\beta_{i,k} = \cos^{-1} \left(\frac{\lambda(\widehat{f}_{beat_{k-1}} - f_i)}{|v|} \right)$$

2. Computing the ML estimate for the beat frequency:

$$\widehat{f}_{beat_k} = \frac{1}{N} \sum_{i=1}^{N} f_i + \frac{|v|}{\lambda N} \sum_{i=1}^{N} \cos \beta_{i,k}$$

where $k = 1, \cdots, 20$ is the iteration index, and the ML estimate is initialized with the average of the observed Doppler-shifted frequencies, $\widehat{f}_{beat_0} = \frac{1}{N} \sum_{i=1}^{N} f_i$.

We show the convergence results for the beat frequency in Figure 5. We observed that the beat frequency estimate converges within a couple of iterations, hence we conservatively chose 20 iterations for the iterative algorithm. A theoretical analysis of convergence of the algorithm is beyond the scope of this paper.

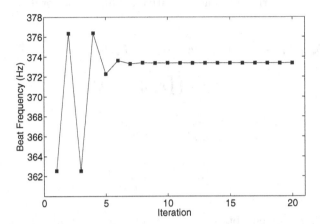

Fig. 5. Convergence results for beat frequency estimate with the maximum likelihood estimation algorithm

4 Implementation

Our wireless sensor platform consists of ExScal motes (XSMs) [11] and a Mobile-Robots Pioneer 3DX [12] robot. All code was written in nesC [13] for the TinyOS operating system [14]. The XSMs use the Texas Instruments CC1000 radio chip [15], and transmit in the 433 MHz band. Note that although the Pioneer comes equipped with an onboard embedded PC, as well as a wide variety of sensors, only the instantaneous velocity, obtained from encoder data, is used, and all computation is performed on the attached mote.

4.1 Implementation Benchmarking

Mobile sensors require a rapid positioning algorithm, otherwise by the time the algorithm completes, the mobile node may be in a completely different location. We therefore provide a timing analysis of our algorithm implementation to demonstrate that its latency is acceptable for mobile sensor navigation. Our method for determining angular separation involves three major steps: (1) signal transmission/reception, (2) sending observed frequencies from the infrastructure nodes to the mobile node, and (3) running the angular separation estimation algorithm. We list the average and maximum latencies for these steps in Table 1. The most unpredictable of these steps is the time it takes the infrastructure nodes to send their observed frequencies to the mobile node. This latency can grow relatively large because the nodes are all attempting to send messages at roughly the same time, resulting in back-off delays. However, even with this unpredictability, we can, on average, obtain angular separation information at a rate of 1.46 Hz, which is sufficient for mobile sensor navigation.

Because we are using resource-constrained sensor nodes, we are also interested in minimizing the memory required to run the algorithm. Our previous

Table 1. Execution time for each step

Step	Average (ms)	Maximum (ms)
Signal transmission	415	417
Routing	242	561
Angular separation algorithm	28	46
Total	**685**	**1024**

approach [6] required the use of two motes on the mobile platform, one hosting the controller, and the other hosting the EKF, leaving little space for the user application. Our current approach requires significantly less memory, using 2.9 kB of RAM and 49.6 kB of program memory (ROM).

5   Evaluation

5.1   Experimental Setup

Our setup consists of 6 XSM nodes, four of which act as stationary receivers and surround the sensing region. Another stationary node is designated the assistant transmitter, and is placed just outside the sensing region. The final mote is attached to the robot. The robot moves around an uneven paved surface in an outdoor environment, mostly free of trees, buildings, and other obstacles. Figure 6 illustrates the experimental setup.

We direct the mobile node to move through the sensing region while transmitting a pure sinusoidal signal. The infrastructure nodes measure the frequency of the signal and report their observations back to the mobile node. At the beginning of each measurement round, the mobile node records its instantaneous

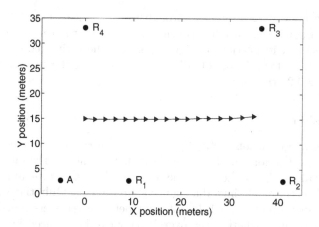

Fig. 6. Experimental setup. Four infrastructure nodes $(R_1 \ldots R_4)$ and the assistant transmitter (A) surround the sensing region. Triangles show the direction of travel of the mobile node at each timestep.

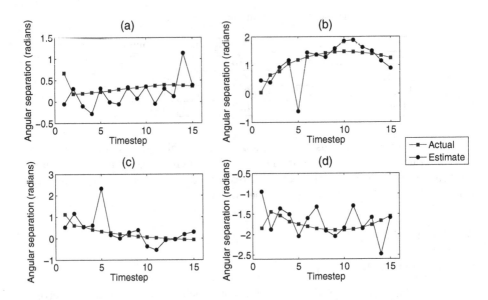

Fig. 7. Actual versus estimated angular separation between receiver nodes (a) R_1 and R_2, (b) R_2 and R_3, (c) R_3 and R_4, and (d) R_4 and R_1, for each measurement as the mobile node traverses the sensing region

velocity, obtained from the wheel encoders. This information is then used to derive the angular separation of the infrastructure nodes. Ground truth is manually measured at each timestep (the time at the beginning of each measurement).

5.2   Experimental Results

As the mobile node makes its way through the sensing region, it periodically computes the angular separation between pairs of infrastructure nodes using the technique presented in Section 3. Figure 7 shows the estimated versus actual angular separations for all pairs of adjacent beacons over the entire course. The average error is 0.28 radian.

5.3   Discussion

The error in our estimation method comes from three main sources: (1) noisy encoder readings, (2) noisy frequency measurements, and (3) the unknown beat frequency. For all receivers, the error due to encoder noise and the unknown beat frequency will be the same, introducing a systematic bias. Only the error due to the frequency measurements will be different between receivers. Further study is needed to determine whether the overall error can be reduced by taking the systematic bias into consideration.

An average angular separation error of 0.28 radian will result in course-grained navigation. Because a significant part of this error comes from the estimation

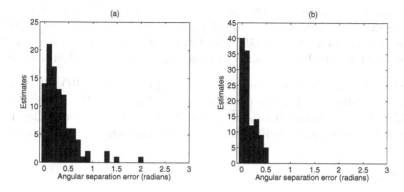

Fig. 8. Error distribution for all adjacent pairs of receivers for each robot position using (a) maximum likelihood estimation, and (b) simulating a known beat frequency

of the beat frequency, we would like to know how this system would respond if we were using radio hardware that transmitted exactly at the desired frequency. With such a system we will still expect measurement noise, whose standard deviation was previously reported to be 0.21 Hz [16], and so we account for this. For the simulation, we use the same infrastructure and mobile node positions as our robot experiment. The average error for this simulation is much lower at 0.14 radian. The error distributions for all pairs of receivers for each robot position are shown in Figure 8.

6 Conclusion

In this paper we presented a feasibility study for determining angular separation using RF Doppler shifts and wheel encoder data in mobile sensor networks. Angular separation between multiple pairs of beacons can be used for navigation, without the need for localization.

Several implementation challenges were encountered while designing this system. Measuring Doppler-shifted frequencies required the use of radio interferometry. Radio hardware limitations caused the actual transmission frequency to be unknown. Because knowledge of the beat frequency was necessary for our algorithm, we used maximum likelihood estimation, however, the accuracy of the results was lower. Experimental results obtained using our method had an average error of 0.28 radian, which will provide course-grained navigation. However, in situations where such navigation is acceptable, our approach is faster, and requires less memory than other RF-based methods (e.g., [16], [17], [6]). This is because our algorithm is distributed, and therefore we expend no time routing data to a base station for analysis. In addition, determining angular separation from Doppler shifts and instantaneous velocity does not require complex statistical tools, such as a Kalman filter, reducing the overall memory footprint of the application.

The goal of our research is to enable mobile sensor navigation using methods that are distributed, require no additional hardware, have low computational complexity, and do not rely on GPS. For future work, we intend to continue this pursuit by examining how we can reduce the angular error for better navigation.

Acknowledgements. This work was supported in part by ARO MURI grant W911NF-06-1-0076, NSF grant CNS-0721604, and NSF CAREER award CNS-0347440. The authors would also like to thank Akos Ledeczi, Janos Sallai, Peter Volgyesi, Metropolitan Nashville Parks and Recreation, and Edwin Warner Park.

References

1. Dantu, K., Rahimi, M., Shah, H., Babel, S., Dhariwal, A., Sukhatme, G.S.: Robomote: enabling mobility in sensor networks. In: The Fourth International Symposium on Information Processing in Sensor Networks, IPSN (2005)
2. Friedman, J., Lee, D.C., Tsigkogiannis, I., Wong, S., Chao, D., Levin, D., Kaisera, W.J., Srivastava, M.B.: Ragobot: A new platform for wireless mobile sensor networks. In: Prasanna, V.K., Iyengar, S.S., Spirakis, P.G., Welsh, M. (eds.) DCOSS 2005. LNCS, vol. 3560, p. 412. Springer, Heidelberg (2005)
3. Wang, G., Cao, G., Porta, T., Zhang, W.: Sensor relocation in mobile sensor networks. In: IEEE INFOCOM (2005)
4. Shah, R., Roy, S., Jain, S., Brunette, W.: Data mules: modeling a three-tier architecture for sparse sensor networks. In: Proceedings of the First IEEE International Workshop on Sensor Network Protocols and Applications (2003)
5. Bekris, K.E., Argyros, A.A., Kavraki, L.E.: Angle-based methods for mobile robot navigation: Reaching the entire plane. In: International Conference on Robotics and Automation (2004)
6. Amundson, I., Koutsoukos, X.D., Sallai, J.: Mobile sensor localization and navigation using RF doppler shifts. In: 1st ACM International Workshop on Mobile Entity Localization and Tracking in GPS-less Environments, MELT (2008)
7. Hightower, J., Borriello, G.: Location systems for ubiquitous computing. IEEE Computer 34(8), 57–66 (2001)
8. Mao, G., Fidan, B., Anderson, B.D.O.: Wireless sensor network localization techniques. Computer Networks 51(10), 2529–2553 (2007)
9. Maróti, M., Kusý, B., Balogh, G., Völgyesi, P., Nádas, A., Molnár, K., Dóra, S., Lédeczi, A.: Radio interferometric geolocation. In: Proc. of ACM SenSys (November 2005)
10. Kay, S.M.: Fundamentals of Statistical Signal Processing, Estimation Theory, vol. I. Prentice-Hall, Englewood Cliffs (1993)
11. Dutta, P., Grimmer, M., Arora, A., Bibyk, S., Culler, D.: Design of a wireless sensor network platform for detecting rare, random, and ephemeral events. In: Proc. of IPSN/SPOTS (April 2005)
12. MobileRobots: Pioneer P3-DX,
    http://www.activrobots.com/ROBOTS/p2dx.html
13. Gay, D., Levis, P., von Behren, R., Welsh, M., Brewer, E., Culler, D.: The nesC language: a holistic approach to networked embedded systems. In: Proc. of Programming Language Design and Implementation (PLDI) (June 2003)

14. Hill, J., Szewczyk, R., Woo, A., Hollar, S., Culler, D., Pister, K.: System architecture directions for networked sensors. In: Proc. of ASPLOS-IX (November 2000)
15. Texas Instruments: CC1000: Single chip very low power RF transceiver (2007), http://focus.ti.com/docs/prod/folders/print/cc1000.html
16. Kusý, B., Lédeczi, A., Koutsoukos, X.: Tracking mobile nodes using RF doppler shifts. In: SenSys 2007: Proceedings of the 5th international conference on Embedded networked sensor systems, pp. 29–42. ACM, New York (2007)
17. Kusý, B., Sallai, J., Balogh, G., Lédeczi, A., Protopopescu, V., Tolliver, J., DeNap, F., Parang, M.: Radio interferometric tracking of mobile wireless nodes. In: Proc. of MobiSys (2007)

Controlling Error Propagation in Mobile-Infrastructure Based Localization

Ying Zhang and Juan Liu

Palo Alto Research Center Inc.
3333 Coyote Hill Rd
Palo Alto, CA 94304, USA
{yzhang,jjliu}@parc.com

Abstract. Iterative localization is one of the common schemes for obtaining locations of unknown sensor nodes when anchor nodes are relatively sparse in the network. The key idea is for a node to localize itself using its anchor neighbors, and then become an anchor for other unknown neighbors. The process continues until all nodes are localized or no nodes left can be localized. The major problem of the iterative localization scheme is that it suffers from the negative effect of error propagation, where sensor noise results in estimation errors which then get accumulated and amplified over localization iterations. This paper proposes a computationally efficient error control mechanism to mitigate the error propagation effect for mobile-infrastructure based localization. In particular, we show how the error can be characterized and controlled in a mobile-assistant localization framework with angle-of-arrival type of sensing modality. Both simulation on a large scale and real experiments on a small scale have been conducted. Results have shown that our error control mechanism achieves comparable location accuracy as global optimization-based localization methods and has the advantage of being much more computationally efficient.

Keywords: Detection and Tracking, Localization, Error Control.

1 Introduction and Motivation

It is important to have the capability to localize objects when a Global Positioning System (GPS) is unavailable (e.g., in an indoor environment) or inaccurate (e.g., in a city center surrounded by high-rise buildings). Much research has been done in localization [1,2,3,4,5,6,7,8,9,10,11,12,13], focusing on developing algorithms to estimate location information based on sensor measurements. Many algorithms are iterative bootstrapping in nature: given nodes with known locations, called anchors, estimate the location for a set of unknown (free) nodes; the estimated nodes are then used to localize others. While this approach is computationally efficient and can easily be distributed, it often suffers from the negative effect of error propagation. Noise in sensor measurements can get amplified due to sensing geometry, and the estimation error can propagate and accumulate during the iterations. This results in poor localization accuracy.

R. Fuller and X.D. Koutsoukos (Eds.): MELT 2009, LNCS 5801, pp. 128–147, 2009.
© Springer-Verlag Berlin Heidelberg 2009

The problem of error propagation and accumulation has been noted in localization literature [14] [15] [16]. Unlike most existing approaches which use certain geometry-related heuristics, our prior work [17] introduces a novel systematic approach, which formalizes the heuristics and performs a quantitative analysis of error characteristics. The basic idea is to document each location estimate with a quality measure and be discretional on which data measurement to use in localization. Recently, similar ideas have been applied to localize mobile nodes [18], where mobile nodes fuse their location estimates based on their relative error estimation. In this paper, we extend the error control mechanism in [17] to mobile-infrastructure based localization [19]. We will investigate where localization error comes from, and how it propagates from nodes to nodes. We will explain in detail how the error control mechanism is devised to manage information with various degrees of reliability. Simulation results show that the error control mechanism significantly improves localization quality, reducing average localization error by a factor of 3 or more, yet it is still computationally efficient, requiring much less computation than optimization-based refinements developed in [19]. Real experiments on a small scale have also been conducted, which verified some of the results obtained from the simulations.

Another body of related work is mobile-assisted localization (MAL). The basic idea is to allow the nodes to move to cover space so that a single node can act like a set of "virtual nodes" to provide more constraints to localization. Most existing MAL-like work [20,21,22,23,24] assumes homogeneous installation, where the mobile nodes are identical to static nodes, and all nodes can both transmit and receive. The mobile-infrastructure-based localization in [19] is different in the sense that the mobile nodes are receivers, powerful in computation capability, while the static nodes are transmitters and do not require any computation.

The contribution of this paper is two-fold. First, it analyzes the source of error propagation for mobile-assistant localization with angle-of-arrival sensing modality, and develops an error estimation mechanism for such a scheme. Second, it verifies that, through both simulation and real experiments, the use of this error control is effective in terms of localization accuracy and computation efficiency.

This paper is organized as follows. Section 2 presents an overview of mobile-infrastructure-based localization developed in [19]. Section 3 analyzes errors from the mobile-infrastructure based localization and describes the error control method which selects neighbors based on location uncertainty. Section 4 evaluates the algorithms given different configuration and noise parameters in simulation and Section 5 presented the results of some small scale experiments in hardware. Section 6 concludes the paper and suggests future directions.

2   Localization with Mobile Infrastructure

In our prior work [19], we have proposed the use of "mobile infrastructure", which uses angle-of-arrival sensors from Ubisense [25]. Two types of nodes are used: ultra-wide band (UWB) transmitters (known as "Ubitags") and UWB receivers (known as "Ubisensors"). Ubisensors determine the angle-of-arrival (AOA) of UWB signals emitted from the Ubitags. Inherently the Ubisensors are more complex than the Ubitags and far more expensive (approximately 40 times in price). The Ubisense system is designed to be a general-purpose location infrastructure. In its intended use, the Ubisensors are permanently installed in a room, and their relative positions are carefully measured and manually input by a skilled technician. Ubitags can be static or moving. The problem with this intended use is the fact that Ubisensor's high price prohibits large scale deployment. To solve this problem, we have proposed the framework of "mobile infrastructure" to make the system more cost effective by taking advantage of the cost asymmetry between Ubisensors and Ubitags. The mobile infrastructure (Fig. 1(a)) consists of a pair of Ubisensors mounted on a mobile cart. In this case, the Ubisensors are considered as infrastructure nodes, moving from place to place, hence the name "mobile infrastructure". In contrast, the Ubitags are deployed in larger quantity, and are considered as ad hoc nodes. Note that the roles of Ubisensors and Ubitags are reversed from the original intended use: mobile sensors are used to localize static tags. This role reversal is cost effective if the unknown locations are widely spread out, spanning multiple sensing ranges, or if the number of mobile objects is much smaller than the number of static objects with unknown locations. The idea of cheap static infrastructure with expensive mobile localizers is applicable to other sensing modalities, such as localizing static RFID tags using a mobile RFID reader, or, localizing visual tags in the environment using a moving camera. In both cases, RFID tags and visual tags are a lot cheaper than RFID readers and cameras, respectively.

The mobile infrastructure constitutes a *sensor frame* with the two AOA sensors in a fixed distance (Fig. 1(b)). Although AOA sensors have six degrees of freedom in general, two of them are fixed in this frame (the pitch and roll angles are fixed to 0). The sensor frame has four degrees of freedom in space: (x_s, y_s, z_s, a_s), where x_s, y_s, z_s are 3D coordinates and a_s is the yaw angle. The sensor frame is mounted vertical, and the roll and pitch angles are both 0. The location of the sensor frame is defined to be the center between two sensors. Without loss of generality, we assume initially the sensor frame is at (0,0,0) and facing $a_s = 0$. The localization problem in this case is to estimate tag locations and the locations (center and yaw) of all subsequent sensor frames based on the AOA measurements. In particular, the measurements are the yaw and pitch angles: $\{\alpha(i, j, l), \beta(i, j, l)\}$, where the index set (i, j, l) denotes the measurement from sensor j to tag i at the l-th sensor frame location. In the absence of noise, the measurement satisfies the following geometry:

$$\begin{cases} (x_i - x_j)\sin(a_j + \alpha_{ij}) = (y_i - y_j)\cos(a_j + \alpha_{ij}) \\ (x_i - x_j)\sin\beta_{ij} = (z_i - z_j)\cos\beta_{ij}\cos(a_j + \alpha_{ij}) \end{cases} \tag{1}$$

where (x_i, y_i, z_i) is tag location, (x_j, y_j, z_j, a_j) is sensor location and α_{ij}, β_{ij} are AOA readings.

For this platform, we have developed a *leapfrog* procedure [19]. It is an iterative procedure. Starting from the initial sensor frame at (0,0,0), it alternates between computation of tag locations and sensor frame locations; from sensors

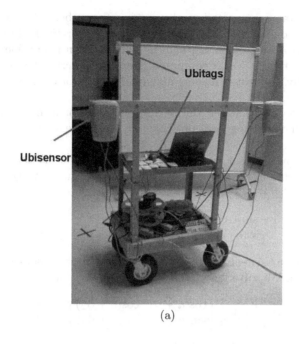

(a)

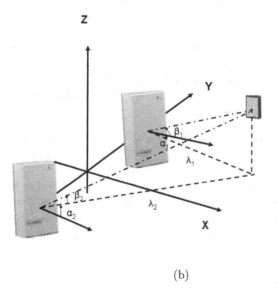

(b)

Fig. 1. (a) Mobile platform and its (b) two-sensor frame

with known location, it estimates tags with unknown locations; and from known tags, it estimates sensor frames with unknown locations. This alternation continues until all the tag locations and frame locations are obtained. In particular, after locating the tags using the current sensor frame, it then selects as the next frame the sensor frame connected to the maximum number of known tags. It computes the location of this new frame, and proceeds iteratively. In the rest of this paper, we call a frame or a tag *known*, if its location is obtained already, and *free* if its location is to be computed. Here we describe the localization computation briefly. Interested readers may refer to [19] for details.

From Sensors to Tags. Given the location of a sensor frame, x, y, z, a, we would like to estimate the tag locations (x_t, y_t, z_t) from the angle measurements $\{\alpha_k, \beta_k\}_{k=1,2}$.

For a tag seen by a sensor frame, let λ_1 and λ_2 be the distances from the tag to the two sensors in the XY plane. From Fig. 1(b), we have the following geometry:

$$\begin{bmatrix} -\sin(\alpha_1) & \sin(\alpha_2) \\ \cos(\alpha_1) & -\cos(\alpha_2) \end{bmatrix} \begin{bmatrix} \lambda_1 \\ \lambda_2 \end{bmatrix} = \begin{bmatrix} d \\ 0 \end{bmatrix}. \tag{2}$$

Here d is the distance between the two sensors in the sensor frame, which is fixed and known a priori. We solve for λ_1 and λ_2. The tag locations are estimated as:

$$\begin{bmatrix} x_t \\ y_t \end{bmatrix} = \begin{bmatrix} x \\ y \end{bmatrix} + \frac{1}{2} \begin{bmatrix} \cos(a + \alpha_1) & \cos(a + \alpha_2) \\ \sin(a + \alpha_1) & \sin(a + \alpha_2) \end{bmatrix} \begin{bmatrix} \lambda_1 \\ \lambda_2 \end{bmatrix}. \tag{3}$$

$$z_t = z + \frac{1}{2}(\lambda_1 \tan(\beta_1) + \lambda_2 \tan(\beta_2)). \tag{4}$$

From Tags to Sensors. If a sensor frame sees multiple known tags, the location of the sensor frame x, y, z, a can be obtained. Let α_{ik} and β_{ik} be yaw and pitch angles from sensor k ($k = 1$ or 2) to tag i, respectively, and let λ_{ik} be the projected distance between tag i and sensor k on the XY plane, which can be computed by (2). It is easy to verify

$$(x_i - x_j)\cos(a) + (y_i - y_j)\sin(a) = \lambda_{ik}\cos(\alpha_{ik}) - \lambda_{jk}\cos(\alpha_{jk})$$
$$(x_i - x_j)\sin(a) - (y_i - y_j)\cos(a) = -\lambda_{ik}\sin(\alpha_{ik}) + \lambda_{jk}\sin(\alpha_{jk})$$

All the λ's are known from Eq.(2). We use an approximate method to solve for a by treating the above as two linear equations of the variables $\cos(a)$ and $\sin(a)$, solving for $\cos(a)$ and $\sin(a)$. Therefore $a = \arctan(\frac{\sin(a)}{\cos(a)})$.

The sensor frame location, i.e., the midpoint between two sensors, computed using tag i is estimated as:

$$\begin{bmatrix} x_s^i \\ y_s^i \end{bmatrix} = \begin{bmatrix} x_i \\ y_i \end{bmatrix} - \frac{1}{2} \begin{bmatrix} \cos(\alpha_{i1} + a) & \cos(\alpha_{i2} + a) \\ \sin(\alpha_{i1} + a) & \sin(\alpha_{i2} + a) \end{bmatrix} \begin{bmatrix} \lambda_{i1} \\ \lambda_{i2} \end{bmatrix}. \tag{5}$$

$$z_s^i = z_i - \frac{1}{2}(\lambda_{i1} \tan(\beta_{i1}) + \lambda_{i2} \tan(\beta_{i2})). \tag{6}$$

For n known tags seen by the frame, the result is the mean of all estimates.

One serious problem of this algorithm is that errors in localization of the previous tags (or frames) will propagate into the localization of the future frames (or tags), accumulating over time. The problem of error propagation and accumulation can be controlled to some degree by optimization-based refinements during or after the leapfrog procedure. It has been shown in [19] that the refinements improve the location accuracy significantly. The problem with optimization-based methods, however, is the computational complexity. It loses the effectiveness for large scale problems, or when the number of variables becomes large. In the rest of this paper, we will focus on how to efficiently mitigate the error propagation and improve localization accuracy.

3   Error Analysis and Error Control

Due to perturbation in sensor observations, localization error may occur. Measurement noise can have an effect on the localization accuracy, and the severity is decided by geometry. For example, if a free node is localized using AOA measurements from known nodes that are far away, then error in angle measurements results in large offset in space. Another pathological case is that the unknown node is localized with a set of nodes that are all collinear with itself. In this case, the location estimate has unbounded error. If this localized node is then used as a known node to localize other nodes, the location error will further affect the accuracy of other nodes. The error could accumulate over localization iterations, and the localization performance can be bad especially in large scale networks.

Our prior work [17] advocates the discrimination of nodes based on a quantitative characterization of their uncertainty. The basic idea is as follows: when a node is localized with respect to its connected known nodes, not all these nodes are equal; certain node may have more reliable location information than others. It is hence preferable to use only reliable known nodes to avoid error propagation. This can be done by keeping track of uncertainty in every step of localization. Any time a location estimate is obtained, the companion estimation for the uncertainty of the location is performed, which formally analyzes how much error is generated in this localization step. This is known as error characterization.

In the rest of this section, we extend this framework to the `Leapfrog` localization procedure and show how uncertainties are characterized and used to mitigate error propagation.

3.1   Leapfrog with Error Control

We defer the problem of error estimation to the next section (Sec. 3.2) and start with a brief overview of the general structure of our error control mechanism. The main idea is to rank nodes (sensor frames and tags) based on the uncertainty of their location information, denoted as e_i for each node i. Only nodes with low uncertainty are used to localize others. This prevents error from propagating. Furthermore, the location update is conditional: a newly estimated location is used to update an old estimate only when the former is believed to have lower uncertainty. This conditional update criterion prevents error from accumulating.

Leapfrog with error control `LeapErrCon` consists of the following components:

(i) *First frame selection*: The first frame serves as reference frame; its location and direction defines the coordinate system. It is also the starting point of the `Leapfrog` procedure. In this first frame, we hope to initialize a substantial number of tags so that later iterations can be "jump-started". Hence we favor choosing a sensor frame with a large amount of connections to tags.

(ii) *Node selection for localization*: For `Leapfrog`, only one frame (the current one) is used for tag localization and all connected tags are used for frame localization. `LeapErrCon` selects a subset of connected known nodes for localization in both cases. The selection is based on predicted errors $\{e_i\}$. `LeapErrCon` ranks nodes with increasing uncertainty e_i and selects the first k, where $3 * mean(\{e_i | i \leq k\}) < e_{k+1}$. This is an empirical criterion which seems to work well.

(iii) *Node update*: In `LeapErrCon`, a tag is localized every time a connected frame is localized. For an unknown tag, both the location and its uncertainty are registered. For a known tag, replace the old registration if the new estimation error is smaller. We further use a forgetting factor update criterion, i.e, $\mathbf{x}_t = (x, y)$, $\mathbf{x}_t = \gamma \mathbf{x}_t^{new} + (1 - \gamma) \mathbf{x}_t^{old}$. This partial update ensures that a noisy measurement will not cause big drift in the location estimate.

(iv) *Next frame selection*: In general, we prefer to first localize tags and sensors with moderate estimation error, and defer noisier estimation problem to later iterations. `LeapErrCon` estimates the total location error for all candidates of the next frame, and selects the one with minimum error.

The pseudo-code is shown in Table 1. Unlike `Leapfrog`, more than one sensor frames may be used to localize a tag and a tag may be localized many times, as long as its location uncertainty decreases.

3.2   Error Estimation for the Leapfrog Procedure

The localization task is to obtain an estimate $(\hat{x}, \hat{y}, \hat{z})$ for any free node (tag or sensor) as a function of other known node locations $\{(x_i, y_i, z_i)\}$ and measurements $\{\alpha_i, \beta_i\}$. Localization error hence comes from two sources: the uncertainty in neighbor positions $\{(x_i, y_i, z_i)\}$ and the uncertainty in each measurement $\{\alpha_i, \beta_i\}$. Assuming we know all the statistics of $\{(x_i, y_i, z_i)\}$ and $\{\alpha_i, \beta_i\}$, can we derive a statistical characterization of the estimate $(\hat{x}, \hat{y}, \hat{z})$? Error characterization attempts to answer this question. In the `Leapfrog` process, error characterization is recursive: the free node derives error characteristics based on neighboring locations and measurement noise. In the next round, this node is used to localize others, hence its error becomes one of the uncertainty sources for the other nodes. Despite the simple formulation, exact error characterization is difficult. We use the same grossly simplifying assumptions as in [17], assuming all noises are Gaussian and the computations are approximately linear. Under these assumptions, error characterization becomes feasible for practical use.

In this section, we analyze error generated from the `Leapfrog` procedure: (a) localizing a tag using known sensor frames (Eq.(3)), and (b) localizing a sensor frame using known tag locations (Eq.(5)). The mobile-infrastructure cart moves

Table 1. Leapfrog with error control `LeapErrCon`; error control parts are in italic

---

Inputs:
  $\alpha_{ijl}, \beta_{ijl}$: angles from tag i to sensor j at frame l
Outputs:
  x_i, y_i, z_i: locations for all tags;
  x_l, y_l, z_l, a_l: locations of sensor frames;
Notations:
  kTs: the set of tags with known locations
  cTs: the set of tags connected to the current known sensor frame
Initialization:
  $l \leftarrow$ the first sensor frame selected using step (i),
  $kTs \leftarrow \emptyset$: no known tags
0. **while there is a new known sensor frame** l:
1. Let cTs be the set of tags connected to frame l;
2. For each tag in cTs:
2.a   *select connected known frames using step (ii)*
2.b   compute the location of the tag
2.c   compute error estimation of the tag using Eq. (7)
2.d   *update both tag location and uncertainty using step (iii)*
3. $kTs \leftarrow kTs \cup cTs$
4. *Let l be the next free sensor frame selected by step (iv)*
5.a *select connected known tags using step (ii)*
5.b compute the location of the frame
5.c compute error estimation of the frame using Eq. (7)
5.d *update both frame location and uncertainty using step (iii)*
6. **end while**

---

in a horizontal plane. The z-dimension is estimated after $(\hat{x}, \hat{y})$ are obtained. Hence from the error-propagation perspective, the error in $\hat{z}$ will not affect error in $(\hat{x}, \hat{y})$. For simplicity, we only characterize error in 2D rather than in 3D. Likewise, the pitch angle β is only used to estimate $\hat{z}$. Hence we ignore this error and only model the error in the yaw angles $\{\alpha\}$.

First, the measurement angle uncertainty in α results in uncertainties in (λ_1, λ_2) (Eq.(2)). Note that λ_1 and λ_2 are the distance from a tag to a pair of sensors, and the solution to Eq.(2) is

$$\begin{bmatrix} \lambda_1 \\ \lambda_2 \end{bmatrix} = \frac{d}{\sin(\alpha_2 - \alpha_1)} \begin{bmatrix} \cos(\alpha_2) \\ \cos(\alpha_1) \end{bmatrix}.$$

The uncertainty due to α_1 and α_2 can be characterized using Taylor expansion, e.g., $\cos(\alpha_1 + \Delta\alpha_1) = \cos(\alpha_1) - \sin(\alpha_1)\Delta\alpha_1$. Assuming that α_1 and α_2 has roughly the same amount of noise, and ignoring error in $\alpha_1 - \alpha_2$, we have the following:

$$cov(\lambda_1, \lambda_2) = \frac{d^2\sigma^2}{\sin^2(\alpha_2 - \alpha_1)} \begin{bmatrix} \sin^2(\alpha_2) & 0 \\ 0 & \sin^2(\alpha_1) \end{bmatrix}$$

where σ^2 is measurement variance of α_1 and α_2. It gives the quantitative classification of the error propagation due to the geometry of the sensor frame and the tag. It is clear that if $\alpha_1 - \alpha_2$ is close to 0, uncertainty from measurements of α to λ is amplified. Intuitively the location estimate error will be big when the two sensors and the tag are collinear, or if the tag is very far away from the sensor frame.

Secondly, the distance pair (λ_1, λ_2) is then used to estimate the tag or sensor position (Eq.(3) or Eq. (5)) via a linear transform B. To characterize the contribution of this term to the overall localization uncertainty, we use the covariance:

$$\Omega^\lambda = B \cdot cov(\lambda_1, \lambda_2) \cdot B'$$

where

$$B = \frac{1}{2} \begin{bmatrix} \cos(\alpha_1 + a) & \cos(\alpha_2 + a) \\ \sin(\alpha_1 + a) & \sin(\alpha_2 + a) \end{bmatrix}.$$

This is assuming that the linear transform B is error-free, hence only the contribution from the distance pair needs to be counted. However, the uncertainty in α_1 and α_2 also directly contributes to B. To take that into account, we now assume that the distance pair (λ_1, λ_2) is error-free and only count the contribution of α_1 and α_2. Note that from Eq.(3) and Eq.(5), we have the location estimate being the sum of three terms: 1. $(x, y)'$, 2. α_1's contribution $\frac{\lambda_1}{2}(\cos(\alpha_1+a), \sin(\alpha_1+a))'$, and 3. α_2's contribution $\frac{\lambda_2}{2}(\cos(\alpha_2+a), \sin(\alpha_2+a))'$. To compute the contribution of these individual terms, we again use Taylor expansion and compute the respective covariance:

$$\Omega^{B_1} = \frac{\lambda_1^2 \sigma^2}{4} \begin{bmatrix} \sin^2(\alpha_1 + a) & 0 \\ 0 & \cos^2(\alpha_1 + a) \end{bmatrix},$$

$$\Omega^{B_2} = \frac{\lambda_2^2 \sigma^2}{4} \begin{bmatrix} \sin^2(\alpha_2 + a) & 0 \\ 0 & \cos^2(\alpha_2 + a) \end{bmatrix}.$$

where Ω^{B_1} and Ω^{B_2} are the contribution from α_1 and α_2 respectively. Let Ω^v be the location uncertainty of the known node v used in location computation, the *location uncertainty* due to both location and measurement errors from the known node is estimated by

$$\Omega = \Omega^v + \Omega^\lambda + \Omega^{B_1} + \Omega^{B_2}. \tag{7}$$

For a set of known nodes used for this computation, the uncertainty is set to be the average of the uncertainty computed from each participated node.

Every time a node location is estimated or refined, this error characterization step is performed. The covariance matrix Ω is computed and stored in the node registry of the newly localized node. It is compact, only of size 2×2, and is updated via light-weight computation. When examining nodes in the neighborhood, we use the trace $\omega = trace(\Omega)$ as the quality measure, or the error estimate e_i for the localization of node i. Larger e_i means weaker confidence in the node i's location, and hence the node should be used more discretionally. Although this error estimate is only an approximation to real error, it does give quantitative information that can guide the use of information.ith X and Y axes as unit scales

4   Simulation Experiments

In this section, we study the effectiveness of the error control method with respect to optimization-based refinements, using three simulated scenarios.

4.1   Simulation Scenarios

We study three scenarios that are most common in indoor localization:

1. *Long Hall* (Fig. 2(a)): tags are distributed on two sides of a long hallway. The mobile platform is moving from one end to the other end, always facing center of the hallway.
2. *Long Walls* (Fig. 2(b)): tags are distributed along two neighboring walls of a room. The mobile platform is moving from one wall to the other, always facing the wall.
3. *High Ceiling* (Fig. 2(c)): tags are distributed on a ceiling. The mobile platform is moving under the ceiling, parallel to the x-axis and facing the center.

The only input scale of the system is the distance d between the two sensors, which is set to 1 without loss of generality[1].

Other constants for simulation are defined as follows. The width of the hallway is 3. The horizontal distances between the neighboring tags or frames are random with standard deviation 0.2. The orientation of the frame has a standard deviation of 0.1 radians. In all cases except the ceiling case, heights of tags are random with zero mean and standard deviation 1, and the height of the frame is fixed to be zero[2]. AOA data are modeled according to the Ubisense system [25]: angle errors are zero mean, with ranges of yaw (α) and pitch (β) angles in $[-1.2, 1.2]$ and $[-1.0, 1.0]$ radians, respectively. Since Ubisensors can see up to 100 meters in line of sight, we assume no distance constraints for sensing.

4.2   Performance Metrics

One of the performance metrics for localization is location accuracy. *Location accuracy* for tag localization is the displacement or error between the actual and estimated tag locations. The error from one tag is the distance between the actual and estimated tag $\rho \triangleq |\mathbf{x} - \hat{\mathbf{x}}|$. For n tags with errors $\{\rho_i\}_{i=1,\cdots,n}$, there are various ways to aggregate the errors, for example, using mean ($\frac{1}{n}\Sigma_i\rho_i$), mean square ($\frac{1}{n}\sqrt{\Sigma_i\rho_i^2}$), or maximum ($\max_i \rho_i$) values. All of these metrics ignore error distributions which are important for many applications. In this paper, we advocate accuracy with 90% confidence, i.e., sorting $\{\rho_i\}_{i=1,\cdots,n}$ in ascending order, and use the $k = \lceil 0.9n \rceil$ entry value as the accuracy metric.

---

[1] Note that we don't have any unit for distances here since it is all relative to what unit we use for the distance between the two sensors. For example, if the distance between the two sensors are d units, all the results will be scaled by d units.

[2] Note that although the algorithm works for varying frame heights, we only simulate frames with fixed height since the mobile platform we have (Fig. 1(a)) can only move on a plane.

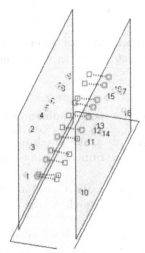

(a) Hallway: tags are on walls, cart moves along hallway, facing center

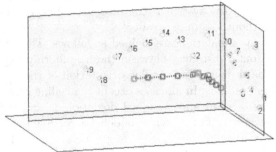

(b) Walls: tags are on walls, cart moves along walls, facing walls

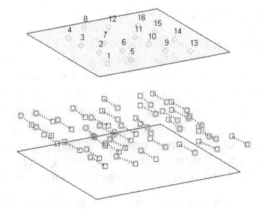

(c) Ceiling: tags are on ceiling, cart moves on floor, facing center of the room

Fig. 2. Three test scenarios: tags are circles and sensors are squares. Each sensor frame is shown by two squares connected by a dotted line, with the left sensor bold.

For each scenario, we generate a total of N random cases, $N = 30$, and again use the accuracy of 90% confidence for representing the accuracy of localization in that scenario. In the rest of this paper, location accuracy or error means location accuracy with 90% confidence.

Another important performance metric is the computation time of obtaining the tag locations. Localization with error control takes more time, but it is a lot more efficient than optimization-based methods.

4.3   Algorithms for Comparisons

There are four variations of the `Leapfrog` procedure, defined by two flags: (1) error control flag (EC), and (2) refinement flag (RF). When EC is off, `Leapfrog` with no error control is used, and when EC is on, `LeapErrCon` (Table 1) is used. When RF is on, optimization-based refinements [19] are applied:

- `Leapfrog`: The `Leapfrog` procedure [19] without refinements.
- `LeapErrCon`: The `Leapfrog` procedure applying the error control mechanism (Table 1).
- `Leapfrog-RF`: `Leapfrog` with refinement, but no error control.
- `LeapErrCon-RF`: `LeapErrCon` with refinement.

4.4   Simulation Results

In this section, we show some simulation results using the four variations of the leapfrog algorithm on the two indoor scenarios. In particular, we are interested in comparing error control mechanisms with optimization-based refinements, from small to large number of nodes, and at different density of tag/cart distributions. In all cases, we generate 30 random data sets and compare location accuracy and localization time using the four variations of the leapfrog procedure. All simulations were run on a PC laptop (Dell Latitude D400, Intel(R) Pentium(R), 1.7GHz, and 1GB RAM) with code written in Matlab.

Hallway. We tested the hallway case using six sets of data with the hallway length scales from 10, 20 to 30, and standard deviations of the AOA noise (σ) of 0.01 or 0.02 radians. Tags were placed liberally on the wall, with an average spacing of 1 (i.e. the same distance as the spacing between the two sensors). The mobile platform (equivalently the sensor frame) was moved along the hallway, stopped randomly with a mean distance of 1 as well, to make measurements.

For 30 random data sets of each case, we also study the error distributions among different variations of the leapfrog algorithm. Fig. 4.4 shows the cumulative error distribution of the hallway case with hallway length 20 for measurement noise σ of (a) 0.01 and (b) 0.02 radians. We see that in both cases, `Leapfrog` does not work well. When the measurement noise is small, all variations except the basic Leapfrog work well, when the noise increases, `LeapErrCon-RF` works the best.

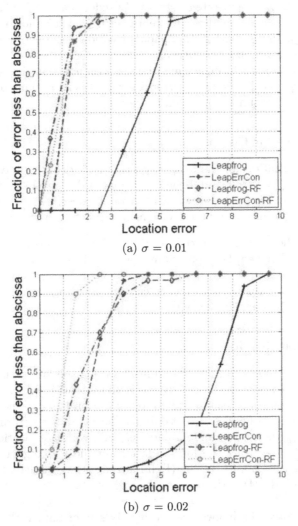

Fig. 3. Cumulative error distribution of the 30 random cases, with hallway length 20, noise 0.01 and 0.02 radians, respectively

Fig. 4(a) shows both the localization time and accuracy for the hallway scenario for measurement noise 0.02 radians. From the result, we can make conclusions that (1) the basic `Leapfrog` procedure does not work well in this case, (2) `LeapErrCon` improves the accuracy significantly, with much less computation time than refinements, (3) `LeapErrCon` works better than refinements for large scales, and (4) `LeapErrCon-RF` has the best accuracy as expected, however, it takes much longer to obtain a result, and the improvement is only marginal. These results clearly demonstrate the effectiveness of `LeapErrCon`.

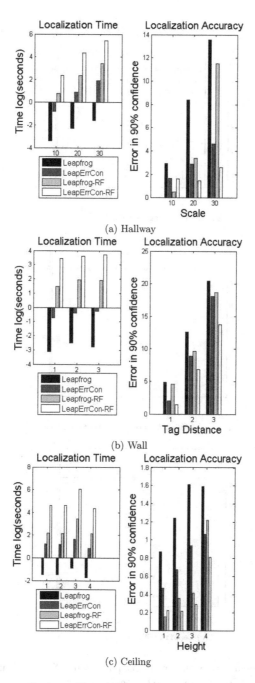

(a) Hallway

(b) Wall

(c) Ceiling

Fig. 4. Performance comparisons (location accuracy and computation time). Note that time is in log scale.

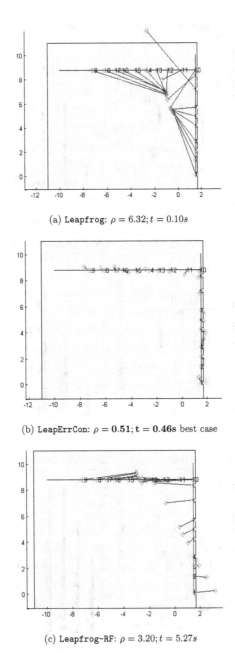

(a) Leapfrog: $\rho = 6.32; t = 0.10s$

(b) LeapErrCon: $\rho = 0.51; t = 0.46s$ best case

(c) Leapfrog-RF: $\rho = 3.20; t = 5.27s$

Fig. 5. Walls tag localization shown in 2D plane with X and Y axes in unit scales. Wall length 10 and AOA variance 0.02. Location accuracy and localization time are shown in ρ and t, respectively for each case. Circles are ground truth and diamonds are estimated locations. Best localization accuracy is shown in bold, which is the case when error control is applied.

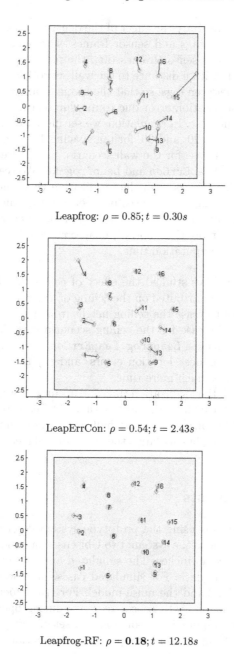

Leapfrog: $\rho = 0.85; t = 0.30s$

LeapErrCon: $\rho = 0.54; t = 2.43s$

Leapfrog-RF: $\rho = \mathbf{0.18}; t = 12.18s$

Fig. 6. Ceiling tag localization shown in 2D with X and Y axes in unit scales. Ceiling height is 2 and $\sigma = 0.02$. Location accuracy and localization time are shown in ρ and t, respectively. Circles are ground truth and diamonds are estimated locations. Best localization accuracy is shown in bold.

Walls. For this case, we study the effect of the density of tag distributions. Like in the hallway case, the tags and sensor frames were spaced with an average distance of 1, 2 or 3. The sensor frames (i.e., the mobile platform) are slightly far away from the wall, with a distance to the wall set to 1.5 times of the distance of the mean spacing between tags, so that the frame at any point can see enough overlapping tags. The location error due to measurement noise would be larger as the tag distances increase. In simulation, we set measurement noise σ to 0.02, and wall length being 10, 20, and 30, for tag spacing 1, 2, and 3, respectively.

Fig. 5 shows one test case for the wall scenario, using three variations of the algorithms: `Leapfrog`, `LeapErrCon` and `Leapfrog-RF`. We see that `LeapErrCon` works better. Fig. 4(b) shows both the localization time and accuracy for the wall scenario, when the mean distances between the tags increase from 1 to 3. From the result, we have the observations that (1) `Leapfrog-RF` is worse than `LeapErrCon`, and takes more time, and (2) `LeapErrCon-RF` has the best accuracy but takes much more computation time.

Ceiling. For this case, we studied the effect of ceiling heights to localization accuracy when tags are distributed on the ceiling of a room. Assuming the frame is at height 0, and we increase the ceiling height from 1 to 4, and $\sigma = 0.02$.

Fig. 6 shows one test case for the ceiling scenario with $H = 2$, using three variations of the algorithms: `Leapfrog`, `LeapErrCon` and `Leapfrog-RF`. We see that (1) `LeapErrCon` reduces location errors, and (2) `Leapfrog-RF` results in best accuracy but takes much more time.

Fig. 4(c) shows both the localization time and accuracy for the ceiling scenario. From the result, we have the observations that (1) LeapErrCon decreases the localization error of Leapfrog by almost half, with less computation time than refinements, (2) when heights are large, the error control mechanism works better than refinements.

5   Real Experiments

We tested our algorithms using the prototype system described in [19]. Figure 1(a) shows the prototype of the system: two Ubisensors are mounted vertically on the poles with distance 40 inches. The simulated and real experiments differed in their source of input data. For simulated cases, data were generated given the tag sensor locations and the noise model. For real experiments, data were generated from continuous AOA sensor readings during operation. In order to get a set of good data inputs corresponding to a set of cart positions, we moved the cart to multiple locations and stopped at each location for 5 to 10 seconds to get stable angle readings.

In this test, we put seven tags on two walls (Fig. 7), and get 3 sets of data. Figure 8 shows the result from the best algorithm using test data 1. To get a sense of the error in these experiments, we use the mean square error (MSE) of the model fitting error. Let $e_{ij\alpha} = 0$ and $e_{ij\beta} = 0$ be the two equations from one tag/sensor pair (i, j) in Eq. 1. We use $\sqrt{\frac{1}{2n}(\Sigma_{i,j}e_{ij\alpha}^2 + \Sigma_{i,j}e_{ij\beta}^2)}$, where n is the

Fig. 7. Experiment setting: tags on two walls

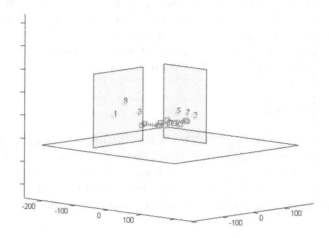

Fig. 8. Result of tag localization: tags on two walls

Table 2. Experimental results: each entry is Mean Square Error for a test using a variation of the algorithm. The best results are shown in bold.

tests	Leapfrog	LeapERCon	Leapfrog-RF	LeapERCon-RF
1	5.57	3.38	2.43	**2.01**
2	4.65	4.42	**2.39**	2.43
3	11.44	9.11	3.67	**3.32**

number of equations, for error estimates. Table 2 shows the results of the 3 sets of data using 4 variations of the algorithm. The best result from each data set is shown in bold. We see that global refinements reduce most of the error, which is consistent with our simulation of small size problems. Also, for two out of three cases, the combinations of error control with global refinements have the best accuracy.

6   Conclusion and Future Work

We have presented an error control mechanism for the leapfrog procedure to localize tags using mobile infrastructure. Simulation showed that the error control scheme significantly improves the localization accuracy and achieves comparable performance as global optimization-based method at only a fraction of the computation cost. Although our real experiments are only in a small scale, we verified that error control would still improve the performance, despite in these cases refinements would work better as we have seen in simulations.

One way to extend this work is to incorporate other kinds of motion data, using inertial sensors or odometry. A wheel-mounted odometric distance measurement could also be combined with the Ubisensor readings to improve the overall system accuracy. Such a system could be smaller with only one sensor. Error control would be even more important in these cases for fusing from different data sources, so as to use the data resulting small error propagation.

Acknowledgement

We would like to thank Dr. Patrick Cheung for helping to build the mobile sensor infrastructure. We also like to thank the regional sales and technical support personal from Ubisense for providing valuable information and assistance. Thanks also to anonymous reviewers and shepherd of this paper Prof. Martin Griss for their constructive comments on the draft of this paper.

References

1. Shang, Y., Ruml, W., Zhang, Y., Fromherz, M.P.: Localization from connectivity in sensor networks. IEEE Trans. on Parallel and Distributed Systems 15(11), 961–974 (2004)
2. Doherty, L., Pister, K.S.J., Ghaoui, L.E.: Convex position estimation in wireless sensor networks. Proceedings of IEEE Infocom. 3, 1655–1663 (2001)
3. Biswas, P., Ye, Y.: Semidefinite programming for ad hoc wireless sensor network localization. In: IPSN (2004) (submitted)
4. Bulusu, N., Heidemann, J., Estrin, D.: Gps-less low cost outdoor localization for very small devices. IEEE Personal Communications Magazine 7(5), 28–34 (2000)
5. Savarese, C., Rabaey, J., Langendoen, K.: Robust positioning algorithms for distributed ad-hoc wireless sensor networks. In: USENIX Technical Annual conference, Monterey, CA (June 2002)

6. Kleinrock, L., Silvester, J.: Optimum transmission radii for packet radio networks or why six is a magic number. In: Proc. IEEE National Telecommunications Conference, pp. 4.3.1–4.3.5 (1978)
7. Costa, J., Patwari, N., Hero, A.: Distributed weighted-multidimentional scaling for node localization in sensor networks. ACM Transactions on Sensor Networks 2(1) (February 2006)
8. Patwari, N., Ash, J.N., Hero, A.O., Moses, R., Correal, N.S.: Locating the nodes. IEEE Signal Processing Magazine 54 (July 2005)
9. LaMarca, A., Chawathe, Y., Consolvo, S., Hightower, J., Smith, I., Scott, J., Sohn, T., Howard, J., Hughes, J., Potter, F., Tabert, J., Powledge, P., Borriello, G., Schilit, B.: Place lab: Device positioning using radio beacons in the wild. In: Gellersen, H.-W., Want, R., Schmidt, A. (eds.) PERVASIVE 2005. LNCS, vol. 3468, pp. 116–133. Springer, Heidelberg (2005)
10. Letchner, J., Fox, D., LaMarca, A.: Large-scale localization from wireless signal strength. In: Proceedings of the National Conference on Artificial Intelligence, AAAI 2005 (2005)
11. Lim, H., Kung, L., Hou, J.C., Luo, H.: Zero-configuration: Robust indoor localization: Theory and experimentation. In: IEEE InfoComm. (2006)
12. Bahl, P., Padmanabhan, V.: An in-building RF-based location and tracking system. In: IEEE InfoComm. (2000)
13. Kaemarungsi, K., Krishnamurthy, P.: Properties of indoor received signal strength for WLAN location fingerprinting. In: IEEE MobiQuitous (2004)
14. Savvides, A., Park, H., Srivastava, M.B.: The n-hop multilateration primitive for node localization problems. Mobile Networks and Applications 8(4), 443–451 (2003)
15. Niculescu, D., Nath, B.: Ad hoc positioning system (APS). GLOBECOM (1), 2926–2931 (2001)
16. Moore, D., Leonard, J., Rus, D., Teller, S.: Robust distributed network localization with noisy range measurements. In: Sensys. (November 2004)
17. Liu, J., Zhang, Y., Zhao, F.: Robust distributed node localization with error control. In: ACM MobiHoc 2006 (2006)
18. Zhang, P., Martonosi, M.: LOCALE: Collaborative Localization Estimation for Sparse Mobile Sensor Networks. In: IEEE/ACM Information Processing in Sensor Networks (2008)
19. Zhang, Y., Partridge, K., Reich, J.: Localizing tags using mobile infrastructure. In: Hightower, J., Schiele, B., Strang, T. (eds.) LoCA 2007. LNCS, vol. 4718, pp. 279–296. Springer, Heidelberg (2007)
20. Pathirana, P.N., Bulusu, N., Savkin, A., Jha, S.: Node localization using mobile robots in delay-tolerant sensor networks. IEEE Transactions on Mobile Computing 4(4) (July/August 2005)
21. Priyantha, N.B., Balakrishnan, H., Demaine, E.D., Teller, S.: Mobile-assisted localization in wireless sensor networks. In: IEEE Conference on Computer Communications, InfoCom 2005 (2005)
22. Taylor, C., Rahimi, A., Bachrach, J.: Simulatenous localization, calibration and tracking in an ad hoc sensor network. In: 5th International Conference on Information Processing in Sensor Networks, IPSN 2006 (2006)
23. Wang, C., Ding, Y., Xiao, L.: Virtual ruler: Mobile beacon based distance measurements for indoor sensor localization. In: The Third International Conference on Mobile Ad-hoc and Sensor Systems, MASS 2006 (2006)
24. Wu, C., Sheng, W., Zhang, Y.: Mobile sensor networks self localization based on multi-dimensional scaling. In: IEEE ICRA 2007 (2007)
25. Ubisense: Ubisense precise real-time location, http://www.ubisense.net

Estimation of Indoor Physical Activity Level Based on Footstep Vibration Signal Measured by MEMS Accelerometer in Smart Home Environments

Heyoung Lee[1], Jung Wook Park[2], and Abdelsalam(Sumi) Helal[3]

Mobile and Pervasive Computing Laboratory
www.icta.ufl.edu
[1] Department of Control and Instrumentation Engineering,
Seoul National University of Technology, Seoul 139-743, Korea
leehy@snut.ac.kr
[2] Department of Electronics and Computer Engineering,
Ajou University, Suwon 443-749, Korea
bl008@ajou.ac.kr
[3] Department of Computer and Information Science and Engineering,
University of Florida, Gainesville, FL 32611, USA
helal@cise.ufl.edu

Abstract. A smart home environment equipped with pervasive net-
worked-sensors enables us to measure and analyze various vital signals
related to personal health. For example, foot stepping, gait pattern, and
posture can be used for assessing the level of activities and health state
among the elderly and disabled people. In this paper, we sense and use
footstep vibration signals measured by floor-mounted, MEMS accelerometers deployed tangent to wall sides, for estimating the level of indoor
physical activity. With growing concern towards obesity in older adults
and disabled people, this paper deals primarily with the estimation of
energy expenditure in human body. It also supports the localization of
footstep sources, extraction of statistical parameters on daily living pattern, and identification of pathological gait pattern. Unlike other sensors such as cameras or microphones, MEMS accelerometer sensor can
measure many biomedical signatures without invoking personal privacy
concerns.

Keywords: Caloric energy expenditure estimation, Indoor activity detection, Localization of footstep source, MEMS accelerometer, Personal
health care, Sensor networks, Smart homes.

1   Introduction

To promote personal health for the elderly and disabled, and to support independent living at reasonable cost, it is agreeable that a home based sensor network

R. Fuller and X.D. Koutsoukos (Eds.): MELT 2009, LNCS 5801, pp. 148–162, 2009.
© Springer-Verlag Berlin Heidelberg 2009

environment that collects various data and vital signs of the residents is a promising approach. In a smart home environment with pervasive intelligence, various vital signals can be measured by sensors such as vision sensor, microphone, electrode, pressure sensor and accelerometer, and can be processed to extract some information on mental and physical states of the resident. Then, the processed data and information can be used to diagnose their health conditions by medical doctors and caregivers.

Chronic diseases, such as diabetes, cardiovascular disease, respiratory disease, obesity, cancer and Alzheimer's disease, are currently leading causes of severe damage among the elderly and disabled people as well as the normal people in the United States. Table 1 shows leading causes of death due to chronic disease. Research on chronic diseases reports that the most common chronic conditions are high blood pressure, high cholesterol, arthritis and respiratory diseases like emphysema [13]. In addition, research on personal healthcare supports that increasing the level of physical activity decreases the risk of onset and developing the chronic illnesses [6][11].

A well-established smart home environment enables us to develop a sensor network-based chronic care management model, which makes it possible to prevent, delay, detect and control chronic diseases. This could be achieved by continuously measuring chronic-related outcomes and by periodically prescribing exercises to control the level of physical activity.

The pattern on the level of outdoor activity fluctuates with environmental changes on temperature, humidity, rainfall, and daylight length since the amount of time spent on outdoor leisure depends on the weather conditions [6][8][12]. The seasonal fluctuation of the outdoor physical activity affects health-related outcomes of the elderly and disabled people [6][8][9]. For example, the studies in [6][8][9][12] on the seasonal variation of blood cholesterol show that the seasonal variation of total physical activity levels in metabolic equivalent (MET)-hour/day is the range of 2.0-2.4 $MET \cdot h \cdot d^{-1}$ in men and women ages 20-70 year [8]. In the northern regions of USA, average total cholesterol peaks in men during the month of December and in women during the month of January when physical activity levels are lower [9]. This suggests that fluctuations on levels of physical activity across seasons may influence health-related outcomes positively or negatively [6][8]. Since the health-related outcomes fluctuates with environmental variations on seasons, controlling level of indoor physical activity

Table 1. Leading causes of death in chronic disease in the Unite States, 2005 [7][10]

Chronic Disease	Number of Deaths	Percentage(%)
Diseases of the heart	652,000	40
Cancer	559,000	34
Stroke	144,000	9
Chronic respiratory disease	131,000	8
Diabetes mellitus	75,000	5
Alzheimer's diseas	72,000	4

adapting to conditions such as weather is conducive to maintaining better health state of the elderly and disable people.

For estimating levels of the indoor physical activity, the vital signs of walking (footsteps), gait pattern and posture are frequently measured via a sensor network-based environment. However, some sensors such as vision sensor and microphone may entail privacy problem despite high security levels of the sensor network itself.

For sensing user location in a smart home environment, we have adapted an unencumbered approach we refer to as smart floor localization and tracking [24]. This approach requires raised flooring (residential grade raised floor) and uses a grid of pressure sensors conditioned for use with psi of human body weight. The smart floor has been deployed at the Gator Tech Smart House (GTSH) [22][23], a real world house that is an assistive environment for R&D in use of pervasive technology for successful aging. The smart floor approach has several drawbacks including high deployment cost (due to required labor for wiring and connecting the sensors), and high complexity of the wiring and connections. The smart floor is also incapable of locating or discerning the locations of more than one user in the house.

In this paper, MEMS accelerometer sensors are considered as a superior technology for location sensing and estimation of indoor physical activities. The MEMS accelerometers are mounted on the floor along side the walls - a constraint that simplifies deployment and that requires no home modifications. The MEMS accelerometers are networked and connected to the smart home computer using the Atlas sensor platform technology [20][21]. The sensors measure vibrations induced by physical activities such as walking, opening and closing doors, washing, eating a meal, cooking, sleeping and watching TV, among other activities. The output of the accelerometers includes various noises caused by not only home appliances such as washing machine and refrigerator with rotator but also vehicles passing by the house. Also, TV and radio may produce induced noise by loud sound pressure. Table 2 shows some noise sources and spectral bands in a smart home environment.

The level of human activity can be manifested by the level of energy expenditure. That is, high level of human activity means that the energy expenditure should be large. For example, a brisk exercise generates forceful large swing motions in the legs, which results in footstep vibration with large magnitude. Therefore, levels of human activity can be estimated from the footstep vibration signal. Table 4 shows energy expenditure for some activities of daily living.

Table 2. Noise source and spectral bands

Source	Spectral Band [Hz]
Vibration by sound pressure from TV and radio	10-22000
Vibration by home appliances with rotator	10-500
Road noise by vehicle	30-60
Impulsive noise induced by closing doors	Larger than 10

With the purpose of developing a smart home-based healthcare system for the elderly and disabled people, this paper deals with the estimation of energy expenditure, localization of footstep sources, extraction of statistical parameters on daily living pattern, and identification of pathological gait pattern, based on connected MEMS sensors, which measures floor vibration induced by footstep on human activity. The purpose of this study is to use a MEMS sensor for localizing footstep source and computing correlation between the level of energy expenditure and the level of floor vibration. In this paper, a footstep vibration is modeled as a seismic signal composed of P-wave and S-wave, and mathematical analysis for localization is conducted based on a least square error method that fits a line of direction, in which 3 projected signals on acceleration to the x-axis, the y-axis and the z-axis are used for line fitting.

Table 5 shows some health-related information and outcomes easily obtainable from footstep localization and tracking. For example, a continuous monitoring based on tracking makes it possible to promptly detect a fall of a frail elderly person.

2   Footstep Signature and Gait Pattern

A person walking on a floor generates a train of impulsive impacts, as the foot hits the floor, which propagates through the floor and produces a footstep vibration as shown in Fig. 1(a), which depends on structural dynamics and material characteristics of the floor in the house [1][2][3]. As shown in Fig. 1(b), a footstep movement is divided into two motion phases, which result in two characteristic spectral bands in the vibration responses of footstep as shown in Fig. 1(c). The footstep force normal to the supporting surface produces a low-banded signal below 500 [Hz] [1][2][3]. On the other hand, the tangential friction force generated by dragging the foot produces a high-banded signal above 1 [kHz] up to ultrasonic spectral range [1][2][3][4][5]. Rhythmic human activities, such as walking, dancing and aerobic introduce a distinct harmonic structure with quasi-periodicity to the resulting vibration responses of footstep. The harmonic structure includes valuable information for studying gait pattern.

The time-frequency representation of footstep vibration signal reveals some information on temporal and spectral variations of footstep and gait pattern. In general, a measured footstep vibration is contaminated by various noises. To clean the measured noisy signal, we can use the variable bandwidth filter, which suppresses noise between peaks without damaging peaks and introducing spurious signal in time-frequency domain. Since walking is one of the most important human activities, studying gait pattern is critical to the monitoring of ambulatory events in the elderly and disabled people [15][17]. Assessing different walking patterns can provide valuable information regarding an individual's mobility, energy expenditure and stability during locomotion [15][[17]. Classification on different walking patterns provides useful information leading to further understanding of both gait pattern and an individual's energy expenditure during daily living [15][17].

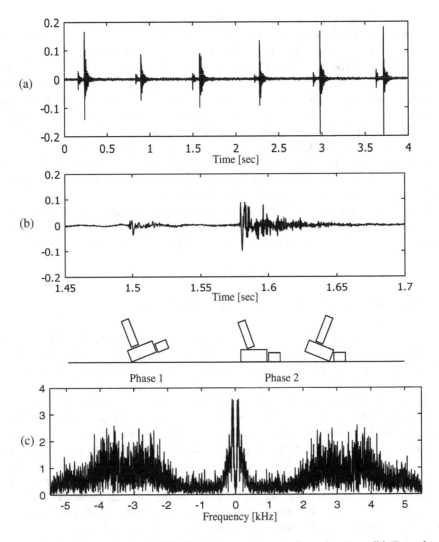

Fig. 1. Human footstep signal. (a) Time-history of vertical acceleration. (b) Two phase motions and corresponding footstep signal. (c) Fourier spectrum.

Patients with diabetes and peripheral neuropathy exhibit gait instability [15][16][17]. Gait unsteadiness has a strong association with depressive symptoms [15][16][17]. Abnormal walkers try to adapt a slower walking speed, shorter stride length, and longer double support time than normal walkers. Similar gait patterns are observed in patients with diabetes and peripheral neuropathy [15][17]. While patients with diabetes adapt a more conservative gait pattern to make them feel more stable, they remain at high risk of falls and injuries during daily activities [15][17].

3   Footstep Source Localization

In health monitoring in smart homes, it is important to obtain information on indoor location of a resident. For example, tracking indoor location is conducive to discriminating a fall of the elderly and disabled if positions of furniture such as couch and bed are known. Also by locating the user, the smart house can better support the user needs by activating certain monitoring and assistive applications that are location and area specific. Table 3 shows some available tools for localizing indoor positions of a subject. In this paper, we consider MEMS accelerometer to localize footstep source.

Table 3. Comparison of indoor localization tools

Source	Information to be Extracted	Difficulty
Speech and sound	2D position	Shade, reverberation, privacy
Active sonar	2D position, orientation	Shade, reverberation
IR LED	3D position, orientation	Shade, cost
UWB	3D position, orientation	Shade, harmfulness
Vision, constellation	3D position, orientation	Complexity, illumination, privacy
Footstep vibration	2D position	Weak signal, space variance

Table 4. Energy expenditure in activity, 160 lbs body mass [14]

Activity	Energy Expenditure [kcal/hr]
Sleeping	70
Lying quietly	80
Sitting	100
Standing at ease	110
Watching TV	110
Conversation	110
Eating meal	110
Strolling	140
Playing violin or piano	140
Housekeeping	150
Walking dog	316
Walking brisk	422

On a floor, vibration signature of human footstep is a kind of seismic wave, which is induced by walking motions. An impact on a floor generates a vibration, which propagates like seismic wave as shown in Fig. 2. Generally, the footstep vibration is composed of two kinds of waves. One is P-wave whose particle motion is parallel to the propagation direction of wave. The other is S-wave whose particle motion is perpendicular to the propagation direction of the wave. A tri-axis MEMS accelerometer can measure projected versions in terms of acceleration on the P-wave and the S-wave with respect to the x-axis, the y-axis and the z-axis respectively. For localization of footstep source, we can use three kinds

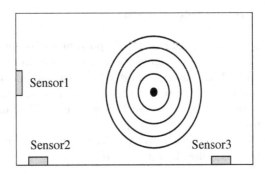

Fig. 2. Propagation of footstep vibration wave

of physical quantities on amplitudes, arrival time differences and directions of particle motion on vibration as follows:

- Based on amplitude
  - Using triangulation
- Based on arrival time difference
  - Using cross correlation between amplitude signals
- Based on direction of particle motion
  - Using P-wave and S-wave vectors

When the decay characteristic on the wave amplitude is known, the trigonometric measure produces estimation on the position of footstep source if we use 3 amplitudes that are measured simultaneously at different places. Also, if we know the propagation speed of the wave, cross correlation of two measured signals at different places produces estimation on the position. Generally, the propagation speed of footstep vibration is larger than 1500 [m/sec]. If the speed of the wave does not depend on the amplitude of the wave, the wave equation that describes the vibration is linear. Therefore, at angular frequency $\omega = 2\pi f$, a zero-state response vector $\mathbf{A}(r, t; \omega) = [A_x(r, t; \omega) \ A_y(r, t; \omega) \ A_z(r, t; \omega)]^T$ for a footstep is represented by

$$\mathbf{A}(r, t; \omega) = \int_{r_s}^{r} \mathbf{H}(r, \sigma, t; \omega)\mathbf{F}(\sigma, t; \omega)d\sigma, \tag{1}$$

where $r = (x, y, z)$ is a position on the space domain and $\mathbf{F}(r, t; \omega)$ is a 3x1 footstep force vector as

$$\mathbf{F}(r, t; \omega) = [F_x(r, t; \omega) \ F_y(r, t; \omega) \ F_z(r, t; \omega)]^T \tag{2}$$

and $\mathbf{H}(r, t; \omega)$ is a 3x3 transition matrix that describes the propagation characteristics of the floor vibration as

$$\mathbf{H}(r, t; \omega) = \begin{bmatrix} H_{11}(r, t; \omega) & H_{12}(r, t; \omega) & H_{13}(r, t; \omega) \\ H_{21}(r, t; \omega) & H_{22}(r, t; \omega) & H_{23}(r, t; \omega) \\ H_{31}(r, t; \omega) & H_{32}(r, t; \omega) & H_{33}(r, t; \omega) \end{bmatrix}. \tag{3}$$

Table 5. Information obtainable from localization

Information and Outcome	Type	Observing Interval
Fall	Strength, tracking	Less than a half hour
Variation on living pattern	Tracking	More than one month
Variation on gait pattern	Strength, tracking	More than one month
Indoor activity level	Strength, tracking	One day
Personal hygiene	Tracking	One day
Habit on eating meals	Tracking	Less than three hours

In general, a house floor is a non-isotropic inhomogeneous elastic medium of seismic waves. For example, a floor is a space-variant dynamic system when some furniture and facilities are on the floor. If a floor has a space-invariant property, that is, a floor is an isotropic homogeneous elastic medium, the matrix $\mathbf{H}(r, t; \omega)$ becomes diagonal. An inverse Fourier transform of $\mathbf{A}(r, t; \omega)$ on the frequency domain produces therefore a time-history of 3x1 vector signals on acceleration.

Figure 3(a) shows a MEMS accelerometer located at $r = (0, 0, 0)$ and a foot-step source at $r_s = (x_s, y_s, z_s)$ on the x-y plane, that is, on the floor. A vibration signature of footstep is composed of P-wave and S-wave. A particle subjected to P-wave moves in the direction that the wave is propagating. P-wave does not generate the vertical acceleration, that is, the z-axis component of acceleration. S-wave moves a particle up and down, or side-to-side, perpendicular to the direction that the wave is propagating. As shown in Fig. 3(b), the footstep source position $r_a = (x_s, y_s, 0)$ can be estimated from the horizontal accelerations, that is, the x-axis and the y-axis components of acceleration, induced by P-wave and S-wave in which directions of particle motion are parallel to the x-y plane.

Let $\mathbf{a}(t)$ be the measured output from a tri-axis MEMS accelerometer as follows:

$$
\begin{aligned}
\mathbf{a}(t) &= [a_x(t)\ a_y(t)\ a_z(t)]^T \\
&= \bar{\mathbf{a}}(t) + \mathbf{n}(t),
\end{aligned}
\tag{4}
$$

where $\bar{\mathbf{a}}(t)$ is a signal on acceleration induced by human activities and $\mathbf{n}(t)$ is a white noise. In general, the P-wave is leading than the S-wave. Also, the P-wave and the S-wave are perpendicular each other.

As shown in Fig. 3, at a time instance t, for the P-wave, we obtain two equations as follows:

$$
\begin{aligned}
y &= (p_{y1}(t)/p_{x1}(t))x \\
&= k_1 x,
\end{aligned}
\tag{5}
$$

and

$$
\begin{aligned}
y &= (p_{y2}(t)/p_{x2}(t))x - (p_{y2}(t)/p_{x2}(t))d \\
&= k_2 x - k_2 d.
\end{aligned}
\tag{6}
$$

Then, the footstep position r_s is represented as follows:

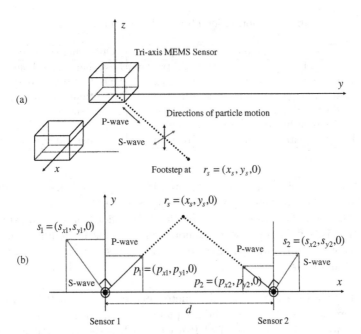

Fig. 3. Configurations for footstep localization. (a) P-wave and S-wave, and corresponding particle motion directions. (b) Localization of footstep source.

$$r_s = (k_2 d/(k_1 - k_2), k_1 k_2 d/(k_1 - k_2), 0). \tag{7}$$

Using the measured time series on the P-wave, the slopes k_1 and k_2 are estimated by the least square approximation on line fitting. That is,

$$k_1 = \frac{E[p_{x1}(t)p_{y1}(t)] - E[p_{x1}(t)]E[p_{x1}(t)]}{E[p_{x1}^2(t)] - (E[p_{x1}(t)])^2} \tag{8}$$

and

$$k_2 = \frac{E[p_{x2}(t)p_{y2}(t)] - E[p_{x2}(t)]E[p_{x2}(t)]}{E[p_{x2}^2(t)] - (E[p_{x2}(t)])^2}, \tag{9}$$

where E means the expectation operation on the measured time series of the P-wave. Also, for the S-wave, we obtain two equations as follows:

$$y = (s_{y1}(t)/s_{x1}(t))x \tag{10}$$

and

$$y = (s_{y2}(t)/s_{x2}(t))x - (s_{y2}(t)/s_{x2}(t))d. \tag{11}$$

As shown in Fig. 3, the S-wave is perpendicular to the P-wave. Using the rotated version of (10) and (11), respectively $\pi/2$[rad] clockwise and $\pi/2$[rad] counterclockwise, we can obtain a representation on the footstep position r_s from the S-wave.

4   Estimation of Energy Expenditure Level

The quantities $\bar{\mathbf{a}}(t)$ and $\mathbf{n}(t)$ in (4) are random variables. Therefore, we obtain that

$$E[\mathbf{a}(t)\mathbf{a}(t)] = E[\bar{\mathbf{a}}(t)\bar{\mathbf{a}}(t)] + E[\mathbf{n}(t)\mathbf{n}(t)] + 2E[\bar{\mathbf{a}}(t)\mathbf{n}(t)], \qquad (12)$$

where E means the expectation operation for a random variable. We assume that $\bar{\mathbf{a}}(t)$ and $\mathbf{n}(t)$ are uncorrelated each other and whose expectation values are equal to zero. Also, we assume that an individual's energy expenditure is proportional to the energy of vibration signal measured by a MEMS accelerometer sensor. Then, from the first assumption, we obtain that

$$E[\bar{\mathbf{a}}(t)\mathbf{n}(t)] = 0.$$

Therefore, (12) is represented by

$$E[\mathbf{a}(t)\mathbf{a}(t)] = E[\bar{\mathbf{a}}(t)\bar{\mathbf{a}}(t)] + E[\mathbf{n}(t)\mathbf{n}(t)]. \qquad (13)$$

Let L_A denote an individual's energy expenditure by activities. In general, the quantity $E[\bar{\mathbf{a}}(t)\bar{\mathbf{a}}(t)]$ is a function of the energy expenditure L_A by human activity, that is,

$$E[\bar{\mathbf{a}}(t)\bar{\mathbf{a}}(t)] = f(L_A). \qquad (14)$$

Then, from the second assumption, we obtain a linear relation as follow:

$$E[\bar{\mathbf{a}}(t)\bar{\mathbf{a}}(t)] = \alpha L_A, \qquad (15)$$

where α is a constant. Combining (13) and (15), we obtain that

$$E[\mathbf{a}(t)\mathbf{a}(t)] = \alpha L_A + E[\mathbf{n}(t)\mathbf{n}(t)]. \qquad (16)$$

As another form, we obtain that

$$\alpha L_A = \frac{1}{\alpha}E[\mathbf{a}(t)\mathbf{a}(t)] - \frac{1}{\alpha}E[\mathbf{n}(t)\mathbf{n}(t)]. \qquad (17)$$

In (16) and (17), $N_0 = E[\mathbf{n}(t)\mathbf{n}(t)]$ represents noise power. If we know the constant α and the noise power $E[\mathbf{n}(t)\mathbf{n}(t)]$, then the energy expenditure L_A can be computed from (17) since $E[\mathbf{a}(t)\mathbf{a}(t)]$ is known. The constant $N_0 = E[\mathbf{n}(t)\mathbf{n}(t)]$ can be estimated from $\mathbf{a}(t) = \bar{\mathbf{a}}(t) + \mathbf{n}(t)$ if $\bar{\mathbf{a}}(t)$ is equal to zero.

5   Variation of Gait Pattern

Negative correlations between age and walking speed, and between age and stride length are observed in the elderly people [18][19]. The relative stance phase duration is correlated positively with age within the elderly people [18][19]. Slow speed is related to low daily activity, reduced muscle power, and diminished

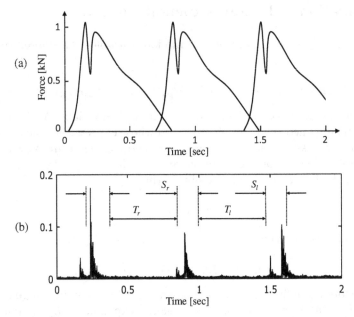

Fig. 4. Human footstep signature. (a) Vertical footstep force. (b) Parameters on one cycle footstep signal.

Table 6. Parameters on gait pattern

Parameter	Description
Duration of stance in sec	S_R, S_L
Cycle duration in sec	$C_D = T_R + S_R + T_L + S_L$
Cycle duty	$C_R = (T_R + S_R)/(T_L + S_L)$
Normalized stride interval	$NS_R = T_R/C_D, NS_L = T_L/C_D)$
Energy of footstep signal	$E_R, E_L, E_T = E_R + E_L$
Normalized energy	$NE_R = E_R/E_T, NE_L = E_L/E_T$
Velocity, footstep/sec	$V_R = 1/(T_R + S_R), V_L = 1/(T_L + S_L)$

balance ability [18][19]. Long stance phase duration and slow speed in the elderly could be an adaptive characteristic in response to impaired balance [18][19].

Figure 4 shows human footstep signature on footstep force and some parameters. As shown in Fig. 1(b), the two phase motions produces two distinct vibration, one is generated by the heel motion normal to the ground and the other is generated by dragging motion tangential to the ground. Parameters, such as duration of stance, footstep cycle and footstep energy, are used for observing variation of gait pattern. From footstep signal (4), we compute an analytic signal to obtain amplitude of (4) as follows:

$$\mathbf{q}(t) = \mathbf{a}(t) + iH[\mathbf{a}(t)], \qquad (18)$$

where H is the Hilbert transform and $\mathbf{q}(t) = [q_x(t)\ q_y(t)\ q_z(t)]^T$. Then, we compute amplitude $|q_z(t)|$, as shown in Fig. 4(b). Table 6 shows parameters,

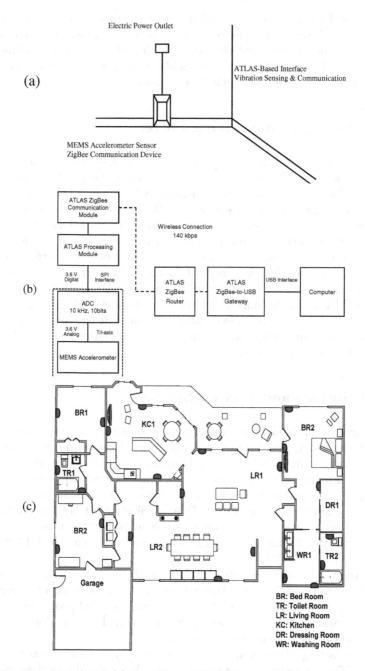

Fig. 5. Experimental setup. (a) A networked MEMS accelerometer attached on the bottom of the wall. (b) Block diagram of the networked MEMS accelerometer. (c) A smart house with network-based MEMS accelerometers connected using the Atlas sensor platform [20][21].

which are used for identifying variation of gait pattern in this paper. Duration of stance, stride interval and energy on right and left foot motion are considered. It can be seen in the amplitude in Fig. 4(b) that the number of walking steps can be easily calculated by using thresholds. The threshold value is chosen as one third of the maximum peak within that frame.

6   Statistics on Daily Living Pattern

Figure 5 shows experimental setup for extraction of statistical parameters on daily living pattern. The relation (17) between energy expenditure and indoor activity level enables us to extract parameters for daily, weekly and monthly charts describing an individual's activities. Based on 24-hour continuous monitoring, generation of statistics on temporal and spatial activity level helps a medical doctor to write exercise prescription of weakness and strength on activities to promote personal health concerns. The difference between exercise prescription recommended by a medical doctor and actual activity level can be estimated based on energy expenditure level and staying time in each space such as bedroom, living room, toilet and kitchen, and on statistics of onset and end of staying interval. Two-dimensional activity map of density on staying time in each living space with temporal information can be constructed for identifying some variation on living pattern.

7   Conclusions and Further Studies

In this paper, a mathematical formulation on localizing footstep source is conducted in which footstep vibration signal is modeled as a seismic wave composed of P-wave and S-wave, where footstep vibration is measured by tri-axis MEMS accelerometers. Since particle motions on P-wave and S-wave include some information of propagation direction, the mathematical formulation enables us to estimate position of footstep source if the number of MEMS sensors for measurement is more than two. To reduce estimation error, the least square error method is used for fitting directional line from footstep source to MEMS sensor location.

Based on MEMS accelerometer, also, we analyze a relation between energy expenditure level and indoor activity level, with the purpose of maintaining personal health conditions among the elderly and disabled people regardless of seasonal variations on weather that affects on personal health-related outcomes such as blood pressure and cholesterol levels.

The long-lasting illness and disability caused by chronic disease decreases quality of life and restricts activities in the elderly and disabled people. The number of steps taken per day is correlated negatively with age in the elderly. Although the elderly are very active, their daily activity is appeared to reduce with age. Slow walking speed is related to daily activity. Long stride length and high speed may be related to muscle power. The main purpose of the system under considering is on estimation of energy expenditure level to promote personal health condition with the help of some networked sensor environments. In the system, the variations on living pattern are measured for on time detection of ambulatory health conditions, based on the statistical parameters extracted from footstep signature and tracking footstep source. Another purpose of the system is to collect continuously some basic bio-signals for transferring to medical doctor.

Acknowledgment. This research is partially funded by an NIH Grant number 5R21DA024294.

References

1. Ekimov, A., Sabatier, J.M.: Vibration and Sound Signatures of Human Footsteps in Buildings. J. Acoust. Soc. Am. 120(2), 762–768 (2006)
2. Ekimov, A., Sabatier, J.M.: Broad Frequency Acoustic Response of Ground/Floor to Human Footsteps. In: Proc. SPIE, vol. 6241 (2006)
3. Houston, K.M., McGaffigan, D.P.: Spectrum Analysis Techniques for Personnel Detection Using Seismic Sensors. In: Proc. SPIE, vol. 5090 (2003)
4. US Patent, Application Number: 60/794,682: Method of Detecting Vibration and Sound Signatures of Human Footsteps, April 25 (2006)
5. Li, X., Logan, R.J., Pastore, R.E.: Perception of Acoustic Source Characteristics: Walking Sounds. J. Acoust. Soc. Am. 90, 3036–3049 (1991)
6. Newman, M.A., Pette, K.K., Storti, K.L., Richardson, C.R., Kuller, L.H., Kriska, A.M.: Monthly Variation in Physical Activity Levels in Postmenopausal Women. Medicine and Science in Sports and Exercise 42(2), 322–327 (2009)
7. American Heart Association, Heart Disease and Stroke Statistics-2007 Update at-a-Glance Version, American Heart Association, Dallas, Texas (2007)
8. Matthwes, C.E., Freedson, P.S., Hebert, J.R., et al.: Seasonal Variation in Household, Occupation, and Leisure Time Physical Activity: Longitudinal Analyses from the Seasonal Variation of Blood Cholesterol Study. AM. J. Epidemiology 153, 172–183 (2001)
9. Ockene, I.S., Chiriboga, D.E., Stanek III, E.J., et al.: Seasonal Variation in Serum Cholesterol Levels. Arch. Intern. Med. 164, 863–870 (2004)
10. US Department of Health and Human Services, Health Behaviors of Adults: United States, 2002-2004. Vital and Health Statistics, Hyattsville (MD), US Department of Health Services, Sept. Available from: US GPO, Washington (2006)
11. US Department of Health and Human Services, Healthy People 2010. 2nd ed.: Understanding and Improving Health and Objectives for Improving Health vol. 2, Washington (DC), US Department of Health Services. Available from: US GPO, Washington (2000)
12. Uitenbroek, D.G.: Seasonal Variation in Leisure Time Physical Activity. Medicine and Science in Sports and Exercise 25, 755–760 (1993)
13. Chronic Care in America: A 21st Century Challenge, a Study of the Robert Wood Johnson Foundation and Partnership for Solutions: Johns Hopkins University, Baltimore, MD for the Robert Wood Johnson Foundation. Chronic Conditions: Making the Case for Ongoing Care (2004)
14. Responsive Environments Group, http://www.media.mit.edu/resenv

15. Kimura, T., Kobayashi, H., Nakagima, E., Hanaoka, M.: Effects of Aging on Gait Patterns in the Healthy Elderly. Anthropology 115, 67–72 (2007)
16. Maynard, F.M., et al.: International Standards for Neurological and Functional Classification of Spinal Cord Injury. Spinal Cord 35, 266–274 (1997)
17. Hardt, D.E.: Determining Muscle Forces in the Leg during Normal Human Walking - An Application and Evaluation of Optimization Methods. J. of Biomechanical Engineering, 72–78 (1978)
18. Ha, S., Han, Y., Hahn, H.: Adaptive Gait Pattern Generation of Biped Robot based on Human's Gait Pattern Analysis. International Journal of Mechanical Systems Science and Engineering (2007)
19. Borghese, A., Bianchi, L., Lacquaniti, F.: Kinematic Determinants of Human Locomotion. Journal of Physiology, 863–879 (1996)
20. King, J., Bose, R., Yang, H., Pickles, S., Helal, A.: Atlas - A Service-Oriented Sensor Platform. In: Proceedings of SenseApp 2006, Tampa, Florida (November 2006)
21. Bose, R., King, J., El-zabadani, H., Pickles, S., Helal, A.: Building Plug-and-Play Smart Homes Using the Atlas Platform. In: Proceedings of the 4th International Conference on Smart Homes and Health Telematic (ICOST), Belfast, the Northern Islands (June 2006)
22. Helal, A., Mann, W., Elzabadani, H., King, J., Kaddourah, Y., Jansen, E.: Gator Tech Smart House: A Programmable Pervasive Space. IEEE Computer magazine, 64–74 (2005)
23. Helal, A., King, J., Zabadani, H., Kaddourah, Y.: Advanced Intelligent Environments. In: Hagrass, H. (ed.) The Gator Tech Smart House: An Assistive Environment for Successful Aging. Springer, Heidelberg (to appear) (2008)
24. Kaddourah, Y., King, J., Helal, A.: Cost-Precision Tradeoffs in Unencumbered Floor-Based Indoor Location Tracking. In: Proceedings of the third International Conference On Smart homes and health Telematic (ICOST), Sherbrooke, Qu?bec, Canada (2005)

Inferring Motion and Location Using WLAN RSSI

Kavitha Muthukrishnan, Berend Jan van der Zwaag, and Paul Havinga

Pervasive Systems Group, University of Twente, Enschede, The Netherlands
{k.muthukrishnan,b.j.vanderzwaag}@utwente.nl
http://ps.ewi.utwente.nl/

Abstract. We present novel algorithms to infer movement by making use of inherent fluctuations in the received signal strengths from existing WLAN infrastructure. We evaluate the performance of the presented algorithms based on classification metrics such as recall and precision using annotated traces obtained over twelve hours effectively from different types of environment and with different access point densities. We show how common deterministic localisation algorithms such as centroid and weighted centroid can improve when a motion model is included. To our knowledge, motion models are normally used only in probabilistic algorithms and such simple deterministic algorithms have not used a motion model in a principled manner. We evaluate the performance of these algorithms also against traces of RSSI data, with and without adding inferred mobility information.

Keywords: Motion inference, Localisation, WLAN, RSSI.

1 Introduction

Ubiquitous computing is emerging as an exciting new paradigm with a goal to provide services anytime anywhere. Context is a critical parameter of ubiquitous computing. Ubiquitous computing applications make use of several technologies to infer different types of user context. The context cue that we are interested in is users' *motion* being either *"moving"* or *"still"* and *location*. The desire of using WLAN infrastructure particularly to derive context is very strong, both from the perspective of the availability of the clients device and that of the infrastructure – nearly all smart phones, PDAs, laptops and many other personal electronic devices have a built-in wireless interface.

Looking at the applicability and usefulness of motion detection, WLAN radio by itself can sense motion, and it can potentially be also part of the sensor ensemble to improve recognition performance. Apart from inferencing activity of the user itself, it has been showcased that such motion inference is useful for efficient radio fingerprinting solutions [2]. Recently movement detection was shown to adaptively switch between passive sniffing and active scanning to allow positioning and to minimise the impact on communications [3]. In this paper, we show yet another use of how it can improve localisation accuracy. The applications that are described above do not necessarily benefit from accurate and complete information about the mobility status. For the purposes described above it is sufficient to know whether the user is moving or not.

R. Fuller and X.D. Koutsoukos (Eds.): MELT 2009, LNCS 5801, pp. 163–182, 2009.
© Springer-Verlag Berlin Heidelberg 2009

This paper examines the results of several motion and location sensing algorithms that operate on RSSI data gathered from existing WLAN infrastructure. The main advantages of the proposed algorithms are: *(i)* deducing user's context without a need for additional hardware, and *(ii)* preserving user privacy, as context inference can be performed locally at the client device.

Contributions: The key contributions of this paper are as follows:

- A detailed characterisation of WLAN RSSI data, exposing a rich set of features both in time and frequency domain to gather mobility information. Our analysis in both temporal and spectral domain, results in a conclusion that "when a device is moving, signal strengths of all heard access points vary much greater compared to when a device is still and the number of detectable samples from access points vary considerably when the device is moving".
- We present novel algorithms to infer movement that makes use of inherent fluctuations in the signal strength. We evaluate the performance of the presented algorithms thoroughly based on classification metrics such as recall and precision from annotated traces (typically groundtruth recorded for every second) obtained over twelve hours effectively from different types of environment and with different access point densities.
- We show how a common deterministic location algorithm such as centroid and its variant can improve its accuracy when a motion model is included. To our knowledge, a motion model is normally used only in probabilistic algorithms and such simple deterministic algorithms have not used a motion model in a principled manner. We evaluate the performance of algorithms against traces of RSSI data collected from different environments, with and without adding mobility information inferred from the mobility detection algorithm.

2   Related Work on Motion Sensing

Randell et al. [9] demonstrated the possibility of distinguishing various states of movement such as walking, climbing and running using a 2D accelerometer. Patterson et al. [6] take the velocity readings from GPS measurements and infer the transportation mode of the user, for instance walking, driving, or taking a bus using a learning model. The model learns the traveller's current mode of transportation as well as his most likely route, in an unsupervised manner. It is implemented using particle filters and is learned using Expectation-Maximisation. The learned model can predict mode transitions, such as boarding a bus at one location and disembarking at another.

Krumm et al. [4] classified a user as either moving or still based on the variance of a temporally short history of signal strength from currently the strongest access point. This classification had many transitions, hence it was smoothened over time with a two-state hidden Markov model resulting in an overall accuracy of 87%.

Anderson et al. [1] use GSM cellular signal strength levels and neighbouring cell information to distinguish movement status. The classification of the signal patterns is performed using a neural network model resulting in an average classification accuracy of 80%. The authors trained the neural network initially and demonstrated a proof of concept by implementing it at run time on a cell phone. However, the initial training did not work in all the environments as signal strength fluctuations were different in different environments. Sohn et al. [10] published a similar technique for detecting the users' motion using signal traces from GSM network. Their motion detection system yields an overall accuracy of 85%. They extracted a set of 7 features to classify the user state as either still, walking, or driving.

Our work on motion detection algorithm is similar to the work on Anderson et al. [1] and Sohn et al. [10], but we look into variation in the WLAN RSSI observed across several access points as opposed to GSM signals.

3   Temporal and Spectral Characterisation of Received Signal Strength (RSSI)

In this section we investigate some of the properties showcased by RSSI, particularly how it changes over time when the user is still and moving, both in static and dynamic environments. By static environment, we mean when the device is placed in a relatively quiet environment (e.g., by logging measurements at off-peak hours) and dynamic environment refers to an area affected by people moving about (e.g., canteen during lunch hours).

Figure 1 shows an example of temporal characteristics – each of the lines represent signal strength received from a specific access point. We can observe from the figure, only some of the access points show a clear distinction between the "still" and "moving" periods – specifically the weaker signals (RSSI < -75 dBm) do not convey any significant difference for both still and moving, hence we have used only the stronger access points for the analysis presented below.

As for whether the variation in the signal strength is influenced more by the changing environment around a static device or by movement of the device itself, Fig. 1 clearly shows that the signal variation is much more prevalent due to the device movement rather than to the dynamics of its environment.

Apart from signal strength fluctuations, we do observe a lot of variations in the number of samples received within a particular observational window (e.g., window of 8

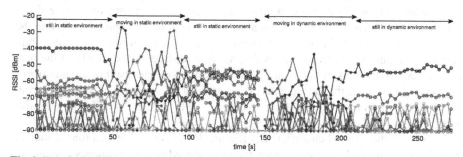

Fig. 1. This figure illustrates two minutes of "still-moving-still" as measured in a static environment and two minutes of "moving-still" measured in a dynamic environment

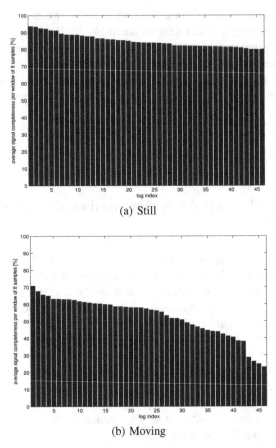

(a) Still

(b) Moving

Fig. 2. Variations in the number of samples received when (a) Still and (b) Moving. Each log (with an average duration of 7–8 minutes) is split into windows of 8 samples and the results are averaged together.

samples) as shown in Figure 2. It is particularly interesting to note that, at a fixed location the number of signal strength samples received from the same access point over a window of reading fairly remains closer to 85% on an average. This is reasonable, as in one scan we typically do not hear all the access points, so one or two missing signal values is still relatively acceptable when the device is still. As opposed to this, in the case of moving, the number of signal strength samples received from the access point varies as the number of access points detectable at a place varies greatly as the user moves. Each of the bins in the Figure 2 represents the average result of the number of samples seen from all the detected access points over all windows of 8 samples from one distinct log. In total for still and moving, we collected over 90 different logs with average duration of log spanning for about 7–8 minutes.

Looking at the spectral characteristics (Figure 3(a)) reveals that as a rule of thumb, the more concentrated the time domain, the more spread out the frequency domain. In particular, if we "squeeze" a function in time, it spreads out in frequency and vice-versa.

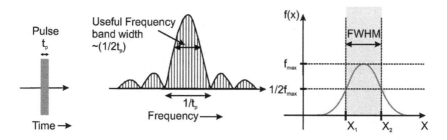

Fig. 3. (a). Schematic representation of a rectangular pulse in time and frequency domain, short duration pulses produces a large bandwidth (b). Full Width Half Maximum, corresponding to peak width at 50% peak height. t_p is the pulse period, and f_{max} is the peak at maximum. X_1 and X_2 are used for calculating FWHM (explained later in Section 4.2).

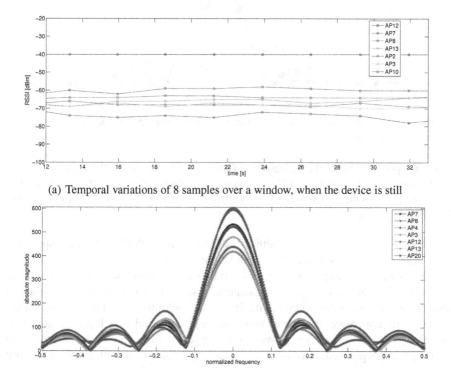

(a) Temporal variations of 8 samples over a window, when the device is still

(b) Spectral variations of 8 samples over a window, with a 512-point FFT when the device is still

Fig. 4. Temporal and Spectral characteristics of a window of 8 samples of the strongest 7 access points for the case of "still". (a) The signal taken is a subset of the signals that are represented in still phase in Figure 1 for time varying between 12–32 seconds, and corresponding FFTs in (b).

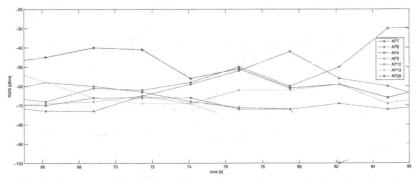

(a) Temporal variations of 8 samples over a window, when the device is moving

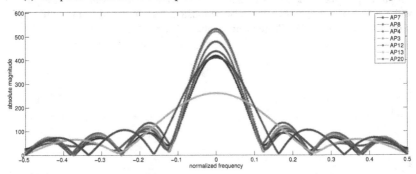

(b) Spectral variations of 8 samples over a window, with a 512-point FFT when the device is moving

Fig. 5. Temporal and Spectral characteristics of a window of 8 samples of the strongest 7 access points for the case of "moving". (a) The signal taken is a subset of the signals that are represented in moving phase in Figure 1 for time varying between 66–86 seconds, and their corresponding FFTs in (b).

Also Figure 3(b) illustrates Full Width at Half Maximum (FWHM) that corresponds to peak width of the FFT signal at 50% peak height.

Figures 4(a) and 5(a) present temporal variations in the signal strength observed over a short window of 8 samples (approximately 20 seconds duration) from the strongest seven heard access points when the device is still and moving and the corresponding frequency domain representation is shown in Figures 4(b) and 5(b). It is evident that although signal strength varies even while the user is still, this variation is reflected in all the heard access points uniformly as there is a well defined peak with a narrow spectral width in the frequency domain from all the access points, despite the fact that there is difference in the Fourier amplitude from each of the heard access point. But when the user is moving, there is no well defined peak from all the access points in the frequency domain indicating that variation in the signal strength happens more often and not in all the heard access points in the same manner. Specifically, we observe the effect of spectral broadening from a significant number of access points when the user is moving, resulting in a wider full width at half maximum. This phenomenon happens mainly due to two reasons: *(i)* the variation in the signal strength is large in case of a moving user and *(ii)* the number of access points detectable varies with distance resulting in too few received samples from the access points. This confirms that both the temporal and spectral analysis lead to the similar conclusions but give a different view of representation. More detailed background information can be found in our earlier work [7].

4 Algorithms for Sensing Motion

In this section we present algorithms for sensing motion. We base the algorithms on the observations presented in the previous section and categorise them into time domain and frequency domain. All presented algorithms are based on "thresholding" applied to a certain metric. We explain in Sec. 4.3 how these thresholds are obtained automatically.

4.1 Time Domain Algorithms

We evaluate four different metrics that we observe in the temporal domain to infer user movement. As opposed to looking at one RSSI value, all the algorithms presented here use RSSI observed over a window of readings (window size typically 8 samples).

AP Visibility. This is the simplest algorithm as it just uses the proportion of the time that RSSI of a particular access point is observed within the observation window. For classifying "moving" or "still" the observed proportion is calculated for each access point and then averaged together. Depending on a (learned) threshold the algorithm detects the state as either moving or still.

Spearman's Rank Correlation Coefficient. We estimate the correlation coefficient using Spearman's Rank Correlation Coefficient (ρ) [11]. The rank correlation coefficient between any two measurements represents how closely the signals are ranked. It takes values between -1 and 1. Values closer to 1 indicate that the measurements are similar and hence the user is still and when the user is moving the values are lower. The algorithm tracks Spearman's ρ between the first and the last measurement in an observation window as a metric to distinguish between moving and still states. Figure 6 presents how the rank correlation coefficient varies when the device is still and moving.

Standard Deviation. This algorithm uses mean standard deviation (SD) over all the heard access points as a metric to distinguish between still and moving states. Within the observation window we measure the SD between the measurements for each detected access point, and use the average SD over all heard access points for inferring the motion status. Figure 7 presents how the average SD varies when the device is still and moving.

Euclidean Distance. This algorithm determines the Euclidean distance between the first and last measurements within an observation window. It is based on the expectation that the average Euclidean distance between WLAN measurements is proportional to the state of the movement. Figure 8 illustrates the average Euclidean distance between WLAN measurements and shows that the average Euclidean distance between WLAN measurements are proportional to the state of the movement.

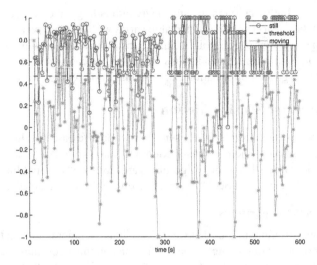

Fig. 6. Spearman's rank correlation coefficient when still and moving, for outdoor (left) and indoor (right) environments. The difference in the rank correlation coefficient remains the same for both outdoor and indoor.

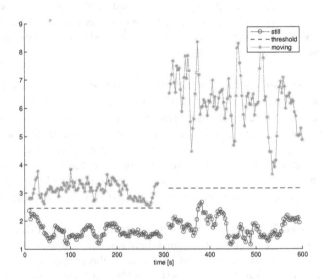

Fig. 7. Mean standard deviation when still and moving, for outdoor (left) and indoor (right) environments. Here we can observe a considerable difference in the SD values between the measurements logged from an outdoor and indoor environment.

4.2   Frequency Domain Algorithms

We now present three novel motion detection algorithms which are inspired by our observations [7] in the frequency spectrum of the RSSI.

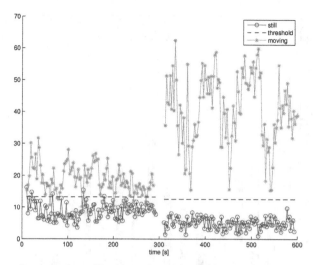

Fig. 8. Euclidean distance when still and moving, for outdoor (left) and indoor (right) environments. Here we can observe a considerable difference in the ED values between the measurements logged from an outdoor and indoor environment.

Full Width at Half Height (FWHM). The RSSI in time series is converted into the frequency domain using Fast Fourier Transformation (FFT). For calculating the FWHM we let the algorithm find the width of the main peak in the FFT signal at 50% of the maximum amplitude, for each of the observed access points. Typically, there is no value at exactly this amplitude, hence we linearly interpolate between the two points nearest to it on either side. For classification, this algorithm uses the FWHM of the main peak of the FFT for a given window of samples and takes the median over all the access points observed in that window.

FWHM Count. This algorithm is very similar to the previous algorithm. It essentially tracks how many access points have a spectral width (i.e., FWHM) that is exceeding a certain threshold within the window of readings. The FWHM is calculated as explained above. Whenever an entry (access point) exceeds the FWHM threshold, the algorithm treats this as an outlier and increments a counter. If the counter exceeds a certain threshold relative to the total number of heard access points within the observation window, the algorithm returns the user state as *moving*, otherwise it returns *still*.

Low-Amplitude-Frequency Count (LAFC). A signal that is not varying much in the time domain has a frequency spectrum with a narrow peak around 0 and very low amplitudes at higher frequencies. In contrast, a signal that significantly varies in the time domain has a broader frequency spectrum, i.e., the peak around 0 is wider and amplitudes at higher frequencies are not as low as for less varying signals.

Based on this observation we use a novel metric, (*low-amplitude-frequency count* or *LAFC*) that distinguishes between "still" and "moving". The algorithm operates on the FFT signal and effectively counts the number of frequencies that have low amplitude

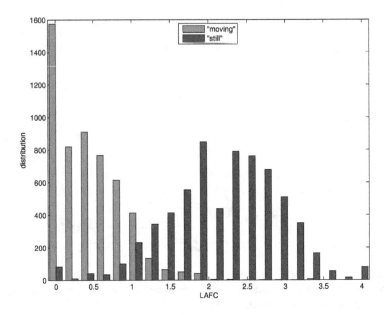

Fig. 9. Distribution of low-amplitude-frequency count (LAFC) for "still" and "moving" classes for a typical training data set (containing more than 12,000 samples in total)

(based on earlier experiments we define " low amplitude" as less than 10% of the maximum amplitude in the FFT to achieve the best results). If this number exceeds a certain threshold, the motion status is set as "moving", otherwise as "still". The LAFC is determined for each heard access point within the observation window and then averaged over all heard access points.

4.3 Threshold Learning

Each of the algorithms described above uses a certain threshold to decide whether the user's device is still or moving. As these thresholds are sensitive to several factors (e.g., environment, hardware, operating system), we use part of our data set for learning the respective thresholds and the remaining part of our data set for determining the classification accuracy using the learned threshold. We use five-fold cross validation, where a data set is partitioned into five folds, and five training and testing iterations are performed. On each iteration, four folds form a training set and one fold is used as a testing set. To illustrate our threshold scheme, let us consider finding the right threshold for the LAFC metric described in Sec. 4.2. The threshold is derived automatically from a training data set (containing more than 12,000 samples in total) using the following method.

1. For each observation window in the training set, the LAFC is calculated.
2. Then, the distributions for both classes ("still" and "moving") are determined. See Figure 9 for an example distribution histogram. If a threshold is applied anywhere on the LAFC axis, typically some of the "moving" observations will lie on the

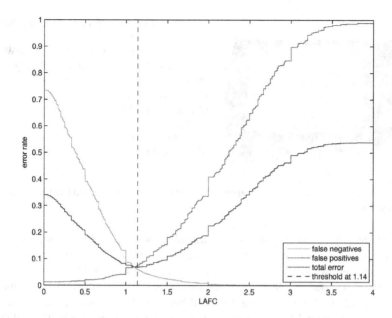

Fig. 10. The amount of false positives ("still" classified as "moving") and false negatives ("moving" classified as "still") as well as their weighted sum as a function of the threshold for the LAFC metric

"still" side (in this case the right hand side) and some of the "still" observations will lie on the "moving" side (in this case the left hand side); these are the *false negatives* and *false positives*.

3. Our method now places the threshold at a position where the weighted sum of false positives and false negatives is minimal. Figure 10 shows the amount of false positives ("still" classified as "moving") and false negatives ("moving" classified as "still") as well as their weighted sum as a function of the threshold. For the LAFC metric, values below (i.e., to the left of) the threshold are classified as "moving" and values above (i.e., to the right of) the threshold as "still".
   We can see that for this particular part of the data set the best threshold is 1.14 yielding a total classification error of about 9% for the training set.
4. We then use this learned threshold to calculate the classification accuracy for the remaining part of the data set (i.e., for the test set). Classification accuracies for all algorithms are reported in the next section.

5 Motion Inference Performance Evaluation

5.1 Data Collection

Two members of our research team collected WLAN network traces, each data collector carried an HP IPAQ pocket PC running a spotter application for recording readings from

Measurment Fri May 25 12:14:44 CEST 2007

25-05-2007 12:14:45 149 Start Sequence:Waaier
25-05-2007 12:14:45 149 Motion:Moving
000136079de0,-90
000cf6164f6c,-90
00116b267fd8,-73
0001e3d43a8d,-53
0001e3da0a55,-90
00147f54a4ff,-74

Fig. 11. Snapshot of (*left*) custom diary application and (*right*) logged ground truth with measured RSSI readings

nearby access points and logging them. Data collectors recorded their mobility activi-ties using a custom diary application running on the PDA that allowed them to indicate whether they were walking, driving, cycling or staying still (refer Figure 11). Data col-lection was performed at common places such as city centre, parking lot, university campus and indoor at the office, canteen and home. In all, the spotter logs contained WLAN traces of about 12 hours duration with annotated ground truth. The unlabeled part of the logs were filtered out in order to measure the accuracies of the presented al-gorithms accurately. Approximately 50% of the logs collected correspond to stationary phase and the remaining 50% correspond to activities performed on the move. Sam-pling the radio environment at approximately 0.4 Hz, the 12 hours of logs correspond to roughly 16,000 samples. The logs also include different access point densities (the least number of access points in the data collected was 0; this happens when no access point is heard during a particular scan, in this case all our algorithms maintain the last inferred motion status until one or more access points are heard again).

5.2   Results and Discussion

We evaluate how accurately the presented *time domain* and *frequency domain* algo-rithms can differentiate between moving and still states. Figure 12 shows a one-to-one comparison of the results obtained from both time and frequency domain algorithms tested against the same data sets.

In order to thoroughly characterise the classification performance, we use the met-rics *precision* and *recall*. Precision for a class is the number of true positives (i.e., the number of items correctly labelled as belonging to the class) divided by the total num-ber of elements labelled as belonging to the class (i.e., the sum of true positives and false positives, which are items incorrectly labelled as belonging to the class). Recall in this context is defined as the number of true positives divided by the total number of elements that actually belong to the class (i.e., the sum of true positives and false negatives, which are items which were not labelled as belonging to that class but should have been). Figure 12 shows the precision and recall of all the 7 algorithms that we discussed in section 4, using five-fold cross validation for the selection of training and test data. The classification results are averaged together for getting the final result.

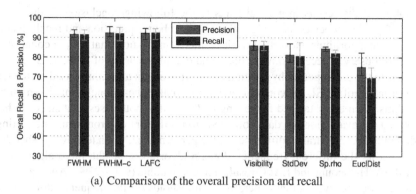

(a) Comparison of the overall precision and recall

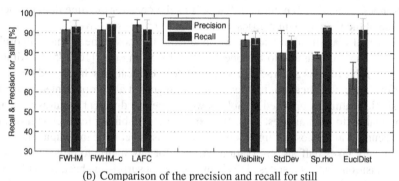

(b) Comparison of the precision and recall for still

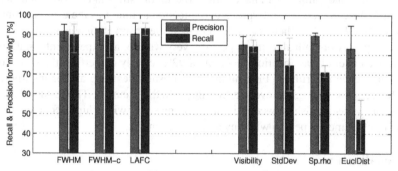

(c) Comparison of the precision and recall for moving

Fig. 12. Performance of motion detection algorithms achieved by 3 frequency domain and 4 time domain algorithms: FWHM, FWHM-count, Low FFT, AP seen, Std dev, Rank, Euc. dist. over 12 hours of WLAN traces collected. The error bars indicate the variations in the accuracy depending on which training and test sets were used in each iteration.

The error bars in Figure 12 indicate the variations in the accuracy depending on which training and test sets were used in each iteration. Further analysis (not reported here) shows that the sensitivity of a particular algorithm does not only depend on the variations in the learned thresholds observed among different folds, but also on the underlying data itself.

Looking at the results in general leads to the following conclusions – the performance of the frequency domain algorithms show a better precision and recall than the time domain algorithms. Nevertheless, it is interesting to note that simple count of the number of observed access points achieves an overall accuracy of 86%. This accuracy is similar to the one reported by Krumm et al. [4] using an algorithm based on the temporal variation of RSSI. Of course it is hard to make one-to-one comparison among algorithms when the data used for testing is different in both cases. But since we have performed the evaluation for data collected from different environments and settings, we do not expect any drastic difference in the performance, when testing on other data.

Comparing the different frequency domain algorithms, all the three algorithms – FWHM, FWHM-count and LAFC results are comparable (91%–92%). Another important aspect we observed is that generally the thresholds for the frequency domain metrics are not as sensitive to external influences as the ones for the time domain metrics and some of the time domain metrics are particularly more sensitive.

The overall classification accuracies obtained with most time domain algorithms (81%–86%) are comparable to the one reported by Sohn et al. [10] (85%). Although Sohn et al. achieved this accuracy for a three state classification scheme and our results are for two state classification, it is interesting to note that all our presented algorithms use a single feature as opposed to Sohn's work where 7 different features were used in combination to train and test data. We expect that combining features will result in better accuracy, at the cost of higher complexity. This is yet to be investigated.

It typically takes half an observation window for any metric to cross its threshold during transitions between states. This is of course not surprising, because halfway the window half of the samples will have a ground truth of "still" and the other half will be "moving". We can therefore interpret a classification at time t as the estimated motion status for time $t - \frac{1}{2}T_w$, where T_w is the length of the observation window.

Our frequency domain algorithms perform very well for all experimental settings by achieving an overall classification accuracy of 92%, clearly outperforming all the other motion inference algorithms. Fine tuning the threshold learning and/or incorporating more features together might even further increase the accuracy. The results show that we are able to distinguish between still and moving states with a high accuracy without having to instrument a person with any additional sensors.

6 Localisation

In this section, we outline the localisation algorithms that operate on RSSI data. The goal of our work is to demonstrate that algorithms which rely on only the location of access points and a coarse estimate of the relative distance to the access points can benefit from adding motion information that we presented above in a principled manner. Many of the probabilistic approaches like particle or Kalman filtering use an inherent motion model to enable localisation and tracking. These algorithms work in a "predictor-corrector" fashion, by weighing the filter model more heavily when the errors in the raw measurement increase, thereby making the final estimates quite accurate. We have used a similar approach but coupled to a simpler centroid algorithm and its variant. This work is an extension of our earlier work [8], which also includes more background and an overview of other related work on WLAN localisation.

Our localisation algorithms are implemented using *filter chains*, which represent a sequence of calculations performed on the RSSI measurements. We rely on the known locations of the access points and other information such as their MAC addresses and their transmit power settings. These access point data could either be managed as a database residing in the network, or it could be a local configuration file in order to minimise the network dependencies and also keeping in view of privacy. Obtaining this information is easier when an accurate database of network access points is already available. In literature, there exist many other ways such as war driving, stumbling to obtain the neighbouring access point coordinates [5].

We use an exponential moving-average filter to reduce the effect caused due to the noise and smoothen the received signal strength for the analysis presented here.

$$\text{RSSI}_{\text{current}} = \alpha \times \text{RSSI}_{\text{prev}} + (1 - \alpha) \times \text{RSSI}_{\text{current}} \tag{1}$$

Equation 1 states that the current RSSI value is a linear aggregate of the previous RSSI value and an independent weighting factor α ($0 \le \alpha \le 1$). The weighting for each older observation decreases exponentially, giving much more priority to recent observations while still not discarding the older observations entirely. We use $\alpha = 0.2$. Automatically determining the optimal α is part of our future work.

The core positioning algorithm is based on a weighted centroid approach. The difference between the normal centroid and weighted centroid is that it introduces variable weights for each access point. Weighted Centroid uses the distance estimates to the strongest access points in relation to each other. This is performed by assigning the location of each of the few strongest access point a weight in the position calculation based on the relative distance between those estimates. Obtaining absolute distances precisely between the access points and the mobile device to be located is harder due to the multi-path reflections that are predominant to indoor environments. Hence algorithms that make use of absolute distances retrieved from WLAN RSSI, such as based on trilateration perform poor, especially when the access points are arranged in a collinear fashion.

We estimate the *relative distances* based on the transmit power of the access points that are available and the RSSI values which typically correspond to the power at the receiving end. Although the distance estimates are not accurate, it will give a cue on which access points are relatively closer and hence to be used in the position estimation. From the RSSI we use motion inference as explained earlier to detect the state of the device as either still or moving. Depending on the state, we use two filter chains:

1. When the motion detection algorithm returns the state of the user as moving, we employ a motion model which smoothens the final location estimates, by preventing any large movements between two different time steps. The motion model filter uses a maximum allowable distance, depending on the users walking speed (say 1.4 m/s) within a stipulated time frame. If the estimated travelled distance exceeds the limit, the location is updated solely based on the motion model.
2. When the user is still, ideally the user's estimated position must remain still at the same point. But since the signal strength varies even at static location, the estimated location often jumps even when the device is still. We therefore use a smaller value for the maximum allowable distance (say 0.2 m) for the static cases and use a history of measurements to average the results together.

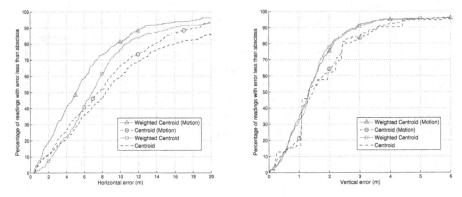

Fig. 13. Cumulative distribution comparing the accuracy of four presented algorithms tested with 20 minutes trace of RSSI measurements

7   Experimental Evaluation

We compare the accuracy of the presented Weighted Centroid with the motion model described above to that of other simple position estimation algorithms with and without adding motion information. In total we have four algorithms to compare: (1) Weighted Centroid with motion model (2) Centroid with motion model (3) Weighted centroid without motion model and (4) Centroid without motion model.

The Centroid algorithm places the user on the geometric centroid of the strongest access points that appeared on the current scan. Weighted centroid as we explained before assigns specific weights to the access points based on the estimated relative distances. Centroid with motion model essentially uses the same principle, but with movement limits depending on the inferred motion status. The metrics we use for the evaluation are the median accuracy, which indicates the accuracy reported by 50% of the readings and the mean horizontal and vertical errors.

7.1   Data Collection

The experiments to assess localisation accuracy were performed in a five-storied university building. Floors 2–5 have a dimension of 106 m × 14.5 m and have a similar layout with a long corridor and many rooms and have four access points per floor that are mounted on the ceiling and are placed in a straight line. The ground floor (refer Fig. 14) has a different layout with a few additional access points covering the northern extension of the building and no access points covering the southern extensions. The transmit powers of the access points are either 50 mW or 30 mW. The measurements were taken from walking along all floors (some twice) from one end to the other, including the stair cases at both ends. Data was recorded as one trace lasting approximately twenty minutes, resulting in about 350 RSSI readings at 0.4 Hz. Since the same data was to be tested with different algorithms, we logged the measurements and all the analyses were done offline.

Table 1. Tracking performance summary. All values shown pertain to the location results of the "walking" data traces collected for about 20 minutes in a five-storied building, covering all the five floors during the measurement period.

Algorithm	Mean hor. error	Mean vert. error	50% conf. level (m) Horizontal	Vertical	75% conf. level (m) Horizontal	Vertical
Centroid	10.77	1.73	8.41	1.30	13.48	2.43
Weighted Centroid	8.53	1.51	6.87	1.34	9.74	1.84
Centroid (with motion)	9.01	1.84	7.49	1.47	12.47	2.42
Weighted Centroid (+ motion)	6.68	1.53	5.14	1.32	8.57	1.98

Ground Truth

Normal map clicking applications are error-prone when used on a small-screen device such as a PDA. Hence, we followed a slightly different approach for registering the ground truth locations. Measurements were logged using the same diary application as used for recording the motion status. At crucial points in the path, such as corners and stairs, the motion status was recorded explicitly so as to log insertion points where we could manually insert the corresponding positions and we used an interpolation script to obtain a ground truth location for each time stamp in the actual measurement log. This worked well and essentially made the comparison easier as we could make a point-to-point comparison between the location estimates and the ground truth.

7.2 Results and Discussion

We evaluate the four algorithms by computing the median accuracy (accuracy reported by 50% of the readings). Table 1 summarises the overall results of the algorithms. Without adding motion, Centroid and Weighted Centroid report median accuracies of 8.41 m and 6.87 m respectively. Adding motion improves these to 7.49 m and 5.14 m respectively. This is because the motion model filter utilises its predicted estimate for the position of the device, in addition to estimates calculated using current RSSI observations, to produce the new estimate (similar to that of a Kalman filter). The cumulative distribution shows even the 75th percentile error reports less than 9 m error for weighted centroid with motion incorporated, given the fact that we have completely avoided the intensive radio-mapping process which is typically used in fingerprinting algorithms. Figure 14 reveals that a considerable portion of the error occurs when the user is at the extreme end of the corridor, as typically the extreme ends have much less access point densities. It is to note that the building has no nearby neighbouring buildings and hence the access points are solely used by the installations in the same building. Figure 14 also shows that the ground floor measurements do not report any location estimates in the southern extension of the floor. This is again due to the unavailability of the access points in that region. Considering the fact that 15% of the readings constitute either stair cases or the southern extension on the ground floor, the reported mean accuracy of the weighted centroid with motion, 6.68 m, is reasonable for a calibration-free approach.

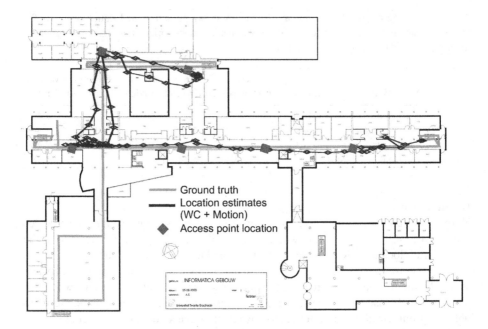

Fig. 14. Access point locations and estimated path overlaid on a floor plan (pertaining to ground floor measurements only). Comparing the trajectory of the ground truth and estimated location, emphasises the error mostly happens on the stairs and towards the extreme end of the corridor.

Table 2. Accuracy of floor estimation, represented on a per-floor basis (percentages)

Algorithm	All floors	Floors 2–5	Accuracy per floor				
			Floor 1	Floor 2	Floor 3	Floor 4	Floor 5
Centroid	70.9	79.4	45.2	72.5	80.3	84.4	83.3
Weighted Centroid	70.0	78.3	45.2	77.5	81.8	75.3	80.0
Centroid (with motion)	72.7	79.5	53.6	71.3	89.4	75.3	86.7
Weighted Centroid (+ motion)	75.1	82.2	53.6	80.0	90.0	75.0	86.0

Floor Identification. This subsection reports how the presented algorithms detects correct floor information. This is very important for many of the applications to identify at which floor a user is present. Table 2 gives the percentage of the measurement time, each algorithm reporting that the user is in the correct floor. For most of the measurements, the algorithm reports that the user is in the correct floor. It is particularly interesting to note that for the measurements in the southern extension on the ground floor the floor error increases. Here we did not have any access points mapped, hence any access points heard at that point were from higher floors, thereby pulling the floor estimates higher by two floors. Table 2 summarises the error in floor estimates reported by all the four algorithms on a per-floor basis. It is clear that the error in the first floor contributes to the maximum error, as all the algorithms consistently show an average of only 50% correct classification. Excluding the first floor measurements show that in principle we are able to identify the correct floor around 82% of the time. The table also shows that there is a modest improvement between the algorithms that use motion information over the ones without a motion model.

8 Conclusions

This paper addresses how to sense motion and location leveraging the existing infrastructure. Based on the thorough characterisation of RSSI measurements we developed a range of motion detection algorithms. We identified a rich set of features that could be gathered based on either temporal or spectral characterisation. Our motion detection algorithms exploiting the frequency domain characteristics report a precision and recall over 90%. It will be interesting to consider the complexity of each of the algorithms to analyse the tradeoff between accuracy and complexity of the presented algorithms. One possibility of extending this work is to use a combined set of features in a machine learning algorithm, to obtain finer accuracy and also explore the possibility of identifying more states like "cycling" or "driving".

We have shown the benefit of combining motion information with location algorithms. A median error of approximately 5 m can be achieved without the use of calibration. We have validated our analysis by testing the algorithms in a typical setup used in many office environments, where access points are arranged linearly. However, these results cannot easily be generalised, as the results are very much dependent on the density and topology of the access points in the test area.

The improvements in the algorithms with motion incorporated, suggest that many of the calibration-intensive fingerprinting algorithms could use such a simple scheme for detecting users in the hallways, and restrict the fingerprints to the rooms, as we envisage that our method might not work as well there. This depends on the access point configuration; if there are access points also distributed along spatially separated axis it will result in considerable improvement because it will allow for better trilateration. In general, if the access points are deployed not only to provide good coverage for communication purposes, but if they are deployed keeping in mind that such infrastructures can be used for positioning purposes, we can expect much more improvement.

When incorporating the motion information we assumed the walking speed of the user is known, usage of other sensors which can actually give us the speed and direction, for instance by using a combination of accelerometers, gyroscopes and magnetometer and incorporating map-matching methods and in combination with probabilistic methods like particle or Kalman filtering might be an suitable venue of future research.

As a general note, we expect that motion information we inferred can also give a cue of what degree of history size must be used for temporal smoothing of the location estimates (i.e., adaptive windowing) – for instance, when the device is moving history size can be set to a smaller value and when the device is still it can be set larger.

References

1. Anderson, I., Muller, H.: Context awareness via GSM signal strength fluctuation. In: 4th International Conference on Pervasive Computing, Late breaking results (May 2006)
2. Bolliger, P., Partridge, K., Chu, M., Langheinrich, M.: Improving location fingerprinting through motion detection and asynchronous interval labeling. In: LoCA 2009. LNCS. Springer, Heidelberg (2009)
3. King, T., Kjærgaard, M.B.: ComPoScan: Adaptive scanning for efficient concurrent communications and positioning with 802.11. In: MobiSys 2008 (2008)
4. Krumm, J., Horvitz, E.: LOCADIO: Inferring motion and location from wi-fi signal strengths. In: Mobiquitous 2004, August 2004, pp. 4–13 (2004)
5. LaMarca, A., Hightower, J., Smith, I., Consolvo, S.: Selfmapping in 802.11 location systems. In: Beigl, M., Intille, S.S., Rekimoto, J., Tokuda, H. (eds.) UbiComp 2005. LNCS, vol. 3660, pp. 87–104. Springer, Heidelberg (2005)
6. Patterson, D.J., Liao, L., Fox, D., Kautz, H.: Inferring high-level behavior from low-level sensors. In: Dey, A.K., Schmidt, A., McCarthy, J.F. (eds.) UbiComp 2003. LNCS, vol. 2864, pp. 73–89. Springer, Heidelberg (2003)
7. Muthukrishnan, K., Lijding, M.E.M., Meratnia, N., Havinga, P.J.M.: Sensing motion using spectral and spatial analysis of WLAN RSSI. In: Kortuem, G., Finney, J., Lea, R., Sundramoorthy, V. (eds.) EuroSSC 2007. LNCS, vol. 4793, pp. 62–76. Springer, Heidelberg (2007)
8. Muthukrishnan, K., Meratnia, N., Lijding, M.E.M., Koprinkov, G.T., Havinga, P.J.M.: WLAN location sharing through a privacy observant architecture. In: COMSWARE, New Delhi, India. IEEE Computer Society, Los Alamitos (2006)
9. Randell, C., Muller, H.: Context awareness by analysing accelerometer data. In: MacIntyre, B., Iannucci, B. (eds.) The Fourth International Symposium on Wearable Computers, pp. 175–176 (2000)
10. Sohn, T., Varshavsky, A., LaMarca, A., Chen, M.Y., Choudhury, T., Smith, I., Consolvo, S., Hightower, J., Griswold, W.G., de Lara, E.: Mobility detection using everyday GSM traces. In: Dourish, P., Friday, A. (eds.) UbiComp 2006. LNCS, vol. 4206, pp. 212–224. Springer, Heidelberg (2006)
11. Weisstein, E.: Spearman rank correlation coefficient. From MathWorld–A Wolfram Web Resource, http://mathworld.wolfram.com/

WASP: An Enhanced Indoor Locationing Algorithm for a Congested Wi-Fi Environment

Hsiuping Lin, Ying Zhang, Martin Griss, and Ilya Landa

Carnegie Mellon Silicon Valley,
NASA Research Park Building 23, Moffett Field, CA 95035, USA
{tony.lin,joy.zhang,martin.griss,ilya.landa}@sv.cmu.edu

Abstract. Accurate and reliable location information is important to many context-aware mobile applications. While the Global Positioning System (GPS) works quite well outside, it is quite problematic for indoor locationing. In this paper, we introduce WASP, an enhanced indoor locationing algorithm. WASP is based on the Redpin algorithm which matches the received Wi-Fi signal with the signals in the training data and uses the position of the closest training data as the user's current location. However, in a congested Wi-Fi environment the Redpin algorithm gets confused because of the unstable radio signals received from too many APs. WASP addresses this issue by voting the right location from more neighboring training examples, weighting Access Points (AP) based on their correlation with a certain location, and automatic filtering of noisy APs. WASP significantly outperform the-state-of-the-art Redpin algorithm. In addition, this paper also reports our findings on how the size of the training data, the physical size of the room and the number of APs affect the accuracy of indoor locationing.

1 Introduction

Location is crucial information to many context-aware mobile applications. Personal navigation, asset tracking, local information search and friend finder all require accurate and reliable location information from mobile devices. While the Global Positioning System (GPS) works quite well outside, it does not work well inside buildings because GPS signals can not penetrate most buildings. Indoor locationing plays an important role in ubiquitous computing and attracts considerable interest in both industry and research. Some systems are designed specifically for indoor locationing but require special infrastructure [1], [8]. With the growth of Wi-Fi networks due to declining prices, increased ubiquity of devices (laptops, cell phones, and other devices using Wi-Fi) and simplified installation of Wi-Fi access points (AP), indoor locationing using wireless LAN (WLAN) is becoming more promising.

There are two major types of WLAN-based indoor locationing approaches: signal propagation model and fingerprinting. In the signal propagation approach, we have to know physical locations of all APs in advance. The received signal strength (RSS) on the mobile device can then be used to estimate the distance

R. Fuller and X.D. Koutsoukos (Eds.): MELT 2009, LNCS 5801, pp. 183–196, 2009.
© Springer-Verlag Berlin Heidelberg 2009

from AP to the mobile device. One can use a multi-lateration algorithm to calculate the the physical location of the mobile device. To reflect the real environment, some signal propagation methods even include wall-attenuation and reflections into the model. In general, the signal propagation approach does not always provide satisfying results because of the Wi-Fi signal fluctuation caused by environmental variations [9].

The fingerprinting approach requires a training database of RSS fingerprints and their corresponding locations. The location of the mobile device is determined by the location of similar fingerprints in the database [2] or from the statistical model derived from the training data [4].

There are many location fingerprinting algorithms. The simplest one is based on the K-nearest neighbor algorithm (KNN). It converts fingerprints into vectors and chooses the K historical fingerprints that are most similar to the testing fingerprint. The location of the testing fingerprint is determined by the majority of its K nearest neighbors. Extending the KNN algorithm, [3] measures not only the contribution of RSSes but also the number of common access points and not-common access points. Another fingerprinting approach is to model the distribution of RSSes at various locations and tries to handle the uncertainty and errors of signal strength measurements .

In this paper, we describe the WASP algorithm, an enhanced indoor locationing algorithm for congested Wi-Fi environments. WASP is a fingerprinting approach and it significantly improves the state-of-the-art indoor locationing algorithm in our experiments.

The rest of this paper is organized as follows. Section 2 reports related work. Section 3 introduces the WASP algorithm and other statistical methods evaluated in this paper. Section 4 describes the dataset, our experimental environment and key results and we conclude the paper with discussion and plans for future work in section 5.

2   Related Work

A reliable and stable interior positioning system (IPS) would be of great benefit to many applications. Considerable research has been performed to determine the indoor location of a mobile user or a mobile device. RADAR, developed by Bahl *et al.*, is an IPS based on Wi-Fi technology [1]. It uses signal strength information gathered at multiple receiver locations by the PC based stations to triangulate the user's coordinates. Paschalidis *et al.* presents an approach that allows a wireless sensor network to determine the physical locations of its nodes by partitioning the wireless sensor network into regions and the localization algorithm identifies the region where a given sensor resides [11].

Most recent research collects RSSes directly on the mobile devices, avoiding the need for extra hardware elements. Li *et al.* compares the trilateration and fingerprinting approaches, including both deterministic methods and probabilistic methods [10]. Brunato *et al.* provide a general comparison of SVM, KNN, Bayesian modeling and multi-layer perceptrons for locationing [4]. Carlotto *et al.* evaluate the proximity of two mobile devices by classifying the degree of similarity of the Wi-Fi scanned data using a statistical Gaussian Mixture Model [5]. Correa *et al.* report experiences using an existing Wi-Fi infrastructure without specialized hardware added to support room-level Wi-Fi location tracking by signature matching, as well as the use of a specialized AP controller [6]. Bolliger proposes the Redpin system, a novel approach that does not require an explicit offline phase but allows users to create and manage the location fingerprints collaboratively [3]. In our work, we started from the open source Redpin system and made substantial enhancements for the congested Wi-Fi environment.

3   Indoor Locationing Algorithms

Our location fingerprinting is based on the assumption that a mobile device will experience a different RSS fingerprint at different locations in the building, and that the variation of the fingerprints seen over time in one location does not vary too much[1]. We collect training data using handsets from several locations in our building. Each training point is a tuple (L, t) of a location label L and the detected RSSes fingerprint $t = (t_1, t_2, ..., t_N)$ where t_i is the RSS received from AP_i. In this section, we first describe several statistical learning algorithms used in our experiments.

3.1   Naive Bayes Classifier

In Naive Bayes approach, we predict a user's location to be L^* if $P(L^*|t)$ is the highest probability of all possible locations:

$$L^* = \arg\max_L P(L|t). \tag{1}$$

By Bayesian theorem, we have

$$L^* = \arg\max_L \frac{P(t|L)P(L)}{P(t)} = \arg\max_L P(t|L)P(L) \tag{2}$$

$P(t)$ is dropped because it does not depend on L. The conditional probability $P(t|L)$ can be estimated by

$$P(t|L) = P(t_1, \dots, t_N|L) = P(t_1|L)P(t_2|L, t_1) \dots (t_N|L, t_1, ...t_{N-1}) \tag{3}$$

With a naive independence assumption that each t_i is conditionally independent of every other t_j for $t_i \neq t_j$, we have

$$P(t|L) = P(t_1|L)P(t_2|L)...P(t_N|L) = \prod_{i=1}^{N} P(t_i|L) \tag{4}$$

---

[1] Of course, the usefulness of this location-based difference and relatively stable fingerprint depends on the placement of the access points, the shape and construction of the building and the sources of noise and fluctuation.

$P(t_i|L)$ can be derived from the historical fingerprints by maximum likelihood estimation (MLE). Thus, the location L can be derived by

$$L^* = \arg\max_L P(L|t) = \arg\max_L P(L) \prod_{i=1}^{N} P(t_i|L) \tag{5}$$

The problem of the naive Bayes method is that the values of the signal strength are not taken into consideration. In other words, $P(t|L)$ is estimated by counting the frequency where s_i is non-zero at location L and only the existence of a set of APs decides the location. To address this issue, Seshadri *et al.* use Bayesian filtering on a sample set derived by Monte-Carlo sampling to compute the location and orientation estimates [12].

3.2   Support Vector Machine (SVM)

The Support Vector Machine is a useful technique for data classification and some research has applied SVM to the indoor locationing problem [4], [7]. A classification task usually involves training and testing data which consist of many data instances. Each instance in the training set contains one target value (class labels) and several attributes (features). The goal of SVM is to produce a model which predicts the target value of data instances in the testing set when given only the attributes. Although SVM is a powerful classification technique, the fluctuations of signals may cause data instance pollution and affect the accuracy.

3.3   K-Nearest Neighbor (KNN)

The K-nearest neighbor algorithm is a method for classifying objects. Given a training data set with labels, KNN classifies a new data point based on the majority of its k-nearest neighbors. For different applications, different distance functions are defined to quantify the "similarity" between the training and testing points. In the simplest case (K=1), the algorithm finds the single closest match and use that fingerprint's location as prediction.

3.3.1   Distance Function

For a testing fingerprint t , the standard KNN algorithm goes through each point (L, s) in the training data and calculates the distance between t and s . The generalized distance is

$$D_q(t, s) = (\sum_{i=1}^{N} |t_i - s_i|^q)^{\frac{1}{q}} \tag{6}$$

Manhattan distance and Euclidean distance are D_1 and D_2 respectively. The unknown location for t is decided by a majority vote from the K shortest distance fingerprints.

KNN is simple to implement and it provides reasonable accuracy. However, one drawback of the standard KNN is that RSSes detected in the same location vary from time to time. The fluctuations likely to cause errors in predicting locations. This can be partially overcome by having multiple fingerprint sets for a given location, taken at different times, assuming that one or other finger print may cover that fluctuation.

3.3.2 Redpin Algorithm - AP Similarity

The Redpin[2] algorithm is a variation of the standard KNN algorithm where the Euclidean distance is augmented with a bonus factor to reward training and testing fingerprints to have common APs and a penalty factor for not-common APs in two fingerprints. Thus, in addition to the signal strength, the number of common access points (NCAP) and the number of not-common access points (NNAP) also contribute to identifying the similarity of two fingerprints. The Redpin algorithm chooses K=1 to decide the best match and works as follows. We define a mapping function $\delta(s)$ as

$$\delta(s) = \begin{cases} 0, \text{ if } s = 0 \\ 1, \text{ if } s \neq 0 \end{cases} \tag{7}$$

NCAP of two fingerprints, t and s , can be expressed as

$$\text{NCAP} = \sum_{i=1}^{N} \delta(t_i)\delta(s_i) \tag{8}$$

NNAP of t and s can be expressed as

$$\text{NNAP} = \sum_{i=1}^{N} \delta(t_i) \oplus \delta(s_i) \tag{9}$$

where $\oplus$ represents the exclusive disjunction. The generalized similarity value of t and s is

$$D(t, s) = \alpha \sum_{i=1}^{N} \delta(t_i)\delta(s_i) - \beta \sum_{i=1}^{N} \delta(t_i) \oplus \delta(s_i) + \gamma \Lambda(t_i, s_i) \tag{10}$$

Λ is a heuristic function defined in the Redpin algorithm which calculates the similarity of t and s based on the signal strengths. The factors α and γ are the bonus-weights for the common APs while β is the penalty-weight for the not-common APs. The key idea behind Redpin is using NCAP and NNAP as bonus-penalty adjustments which reduces the impact of signal fluctuations.

3.3.3 Weighted AP Similarity

To further reduce the impact of signal fluctuations, we observe that the visibility of the APs at one location is not always the same because the environmental variations cause significant Wi-Fi signal fluctuations in the same location over time, especially inside a large building with sparse APs. Intuitively, APs with higher visibility at a location L should be weighted more in determining whether

---

[2] The open source Redpin can be found at http://www.redpin.org

a fingerprint is located at L. In this paper, we use the correlation between APs and locations as the weight for each AP. We use the Point-wise Mutual Information (PMI) as the correlation measurement. PMI is defined as

$$I(L; AP) = log \frac{P(L, AP)}{P(L)P(AP)} \tag{11}$$

The higher the $I(L; AP)$ value, the more likely L is associated with AP. From the historical fingerprints in the database, we can calculate the $I(L; AP)$ value of each location L and AP pairs. We normalize the PMI value to be between 0 (least correlated) and 1 (most correlated). PMI values are applied as weighting modifiers to the bonus of each common AP (CAP) and the penalty of each not-common AP (NAP). The weighted similarity value of the measured fingerprint t and a historical fingerprint s located at L is

$$D(t, s) = \alpha \sum_{i=1}^{N} \delta(t_i)\delta(s_i)I(L; AP_i) - \beta \sum_{i=1}^{N} \delta(t_i)\oplus\delta(s_i)I(L; AP_i) + \gamma\Lambda(t_i, s_i) \tag{12}$$

3.3.4   Noise Filter (NF)

Extending the idea of weighting APs based on their visibility at each location, we can filter out some APs from one location if they are irrelevant to this location since not all APs have the same contribution to one location. We treat those APs that occur less than the average frequency as irrelevant APs. The remaining APs of the fingerprints are considered as "relevant APs". The average frequency of APs to a location is calculated as

$$\bar{C}(L, AP) = \frac{1}{N} \sum_{i=1}^{N} C(L, AP_i) \tag{13}$$

where $C(L, AP_i)$ is the frequency of AP_i visible from the location L in the training data. NF is then a mapping function which maps the fingerprint s to s', where

$$s_i' = \begin{cases} 0, & \text{if } C(L, AP_i) < \bar{C}(L, AP) \\ s_i, & \text{if } C(L, AP_i) \geq \bar{C}(L, AP) \end{cases} \tag{14}$$

4   Experiment

We test and compare different indoor locationing algorithms in a two-floor campus building with a congested Wi-Fi environment. The WLAN in this building is composed of 16 APs, including seven 3-COM APs, six Motorola APs and three external APs. The fingerprints are collected from the second floor, an area of 60mx15m with a 15mx12m lounge. The floor plans and the locations of APs are shown in Fig. 1.

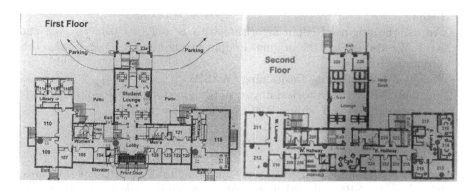

Fig. 1. The floor plans. (The dots are the locations of APs).

4.1 Data Collecting

We selected nine public rooms from the second floor and collected 1,002 unique fingerprints[3] from these locations over a period of seven days using Nokia N95 smart phones. For each room, we collected at least 100 fingerprints to ensure that every location has enough training data (Table 1). Although the size of the rooms is different, instead of measuring the mobile users by physical distance, we find that room-level location information is useful enough for most applications. Therefore, the following experiments are based on room-level location detection.

Table 1. The fingerprint distribution for each room

Rm211	Rm212	W. Lounge	W. Hallway	Cafe	Lounge	E. Hallway	E. Lounge	Rm213
150	100	125	101	101	100	100	124	101

4.2 Experiment Setup and Evaluation

We use stratified 10-fold cross-validation to evaluate the accuracies of different indoor locationing algorithms. To measure the confidence interval of the accuracy, we use the repeated random sub-sampling validation where we repeat the process, randomly choosing 90% of all fingerprints as training data and the remaining 10% as testing data for 100 times. The algorithms we evaluate include:

- Naive Bayes Classifier (NBC)
- Support Vector Machine (SVM)
  We use LIBSVM[4] to infer the locations of the measured fingerprints [13].
- K-Nearest Neighbor (KNN)
  We choose K=5 in our KNN implementation.

---

[3] A text file containing all fingerprints can be found at
http://mlt.sv.cmu.edu/WASP/data.csv

[4] LIBSVM software is available at http://www.csie.ntu.edu.tw/~cjlin/libsvm/

- Redpin

  We use the original Redpin algorithm with $\alpha = 1$, $\beta = 0.4$, $\gamma = 0.2$ and K=1. We also extend the original Redpin by choosing K=5 and name this variation "Redpin5".
- Weighted AP Similarity Positioning (WASP)

  We apply NF to the historical fingerprints and extend the Redpin algorithm by choosing K=5 and adding PMI to weight different APs.

4.3 Result

The accuracies of different algorithms are shown in Table 2. NBC has the lowest accuracy because NBC only calculates the existence of a particular set of APs without considering the signal strengths. When two rooms are quite close to each other, the detected fingerprints are too similar to accurately discriminate two separate rooms. SVM and KNN have similar accuracy because they both use the signal strength information to separate fingerprints from different locations. The Redpin algorithm has better performance than KNN because it reduces the signal fluctuations by using NCAP and NNAP as bonus-penalty adjustments. The WASP algorithm we propose in this paper outperforms the original Redpin by 9% and the 95% confidence interval of the improvement is [0%, 17%], which is statistically significant.

Table 2. Accuracy of each algorithm

	NBC	SVM	KNN	Redpin	Redpin5	**WASP**
Accuracy	61%	80%	79%	81%	86%	**87%**
Confidence interval(95%)	54%-68%	75%-86%	71%-85%	76%-88%	80%-92%	**86%-96%**

Since KNN, Redpin and WASP are all instance-based learning algorithms, we compare their accuracy using different numbers of nearest neighbors (K). The result is shown in Table 3. Increasing the number of nearest neighbors leads to higher accuracy. However, we do not see any major improvement after K reaches 5. Redpin+PMI consistently improves over the original Redpin by around 1%. Though not obvious, the correlation between locations and APs does contribute to the accuracy. More research on alternative statistical methods for the correlation is planned for the future.

To see if NF can successfully reduce the impact of signal noise in the congested Wi-Fi environment for all algorithms, we apply NF to the training dataset and

Table 3. Accuracy of KNN, Redpin and WASP (K from 1 to 10)

	K=1	K=2	K=3	K=4	**K=5**	K=6	K=7	K=8	K=9	K=10
KNN	78%	78%	79%	80%	**79%**	79%	78%	79%	78%	78%
Redpin	81%	81%	83%	85%	**86%**	86%	85%	86%	86%	85%
Redpin(PMI)	82%	82%	84%	85%	**87%**	87%	87%	87%	86%	86%
WASP	88%	88%	90%	90%	**90%**	90%	90%	91%	90%	90%

Table 4. Accuracy of each algorithm without and with NF

Accuracy	NBC	SVM	KNN	Redpin	WASP
W/O NF	61%	80%	79%	81%	87%
With NF	**64%**	**86%**	**88%**	**88%**	**90%**

Table 5. The maximum number of original APs and relevant APs from each location

	Rm 211	Rm 212	W. Lounge	W. Hallway	Cafe	Lounge	E. Hallway	E. Lounge	Rm 213
Original APs	11	12	14	13	15	12	14	14	6
Relevant APs	**6**	**7**	**7**	**8**	**5**	**5**	**9**	**5**	**6**
Invisible APs	5	4	2	3	1	4	2	2	10

run the same experiment. The result is shown in Table 4. All algorithms benefit from NF and the accuracies are improved by 3% to 9%. To better understand the relevant APs, we list the maximum number of the original APs, the relevant APs and the invisible APs in the fingerprints of each location in Table 5 and the visibility of APs from each location in Table 6.

Table 6. The visibility of APs from each room (The relevant APs of each room are bold)

	Rm 211	Rm 212	W Lounge	W Hallway	Cafe	Lounge	E Hallway	E Lounge	Rm 213
AP1	0%	1%	5%	26%	52%	**61%**	**52%**	3%	0%
AP2	16%	1%	**42%**	**69%**	99%	**100%**	**79%**	18%	0%
AP3	0%	0%	1%	34%	**95%**	**100%**	50%	2%	0%
AP4	**65%**	**98%**	**99%**	**100%**	99%	**95%**	**100%**	84%	**61%**
AP5	8%	0%	1%	0%	0%	18%	4%	12%	0%
AP6	**43%**	**62%**	**82%**	**71%**	**83%**	44%	**66%**	15%	0%
AP7	**57%**	**94%**	**89%**	**77%**	**78%**	8%	62%	44%	**38%**
AP8	1%	23%	13%	46%	39%	1%	49%	85%	**96%**
AP9	5%	**39%**	34%	**60%**	58%	6%	62%	94%	81%
AP10	**66%**	**98%**	**100%**	87%	36%	10%	34%	2%	0%
AP11	**65%**	**97%**	**99%**	**85%**	37%	0%	40%	15%	0%
AP12	1%	6%	2%	18%	23%	0%	17%	28%	**34%**
AP13	0%	1%	1%	28%	26%	0%	**78%**	97%	98%
AP14	**65%**	**95%**	**97%**	**78%**	10%	0%	19%	1%	0%
AP15	0%	0%	0%	0%	1%	2%	0%	0%	0%
AP16	0%	0%	0%	0%	8%	3%	0%	0%	0%

4.4   Granularity of Rooms

In addition to the overall accuracy, we also want to know the room-level accuracy. The room-level accuracy is shown in Fig. 2. Surprisingly even though NBC has overall the worst accuracy, it has the highest accuracy in Room213. The most

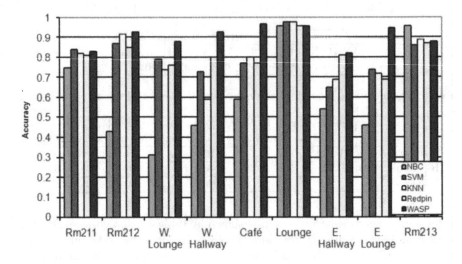

Fig. 2. Room-level accuracy

plausible explanation is that there are ten invisible APs in Room213 which is the highest value among all rooms (see Tables 5, 6). This makes the fingerprint in Room213 the most distinguishable AP set from which NBC can identify its location easily. Another interesting finding is that all algorithms have very high accuracy for the Lounge. Our hypothesis is that the Lounge is the largest room so the estimate error is less significant.

To prove this hypothesis, we create a virtual floor plan by combining adjacent rooms into a larger virtual room. For example, we merge Room211, Room212 into one virtual room (Room 211-212) and a testing fingerprint from Room211 is treated the same as Room212. The accuracy of each virtual room is shown in Fig. 3. For finer-grained locations, the WASP algorithm is the most accurate.

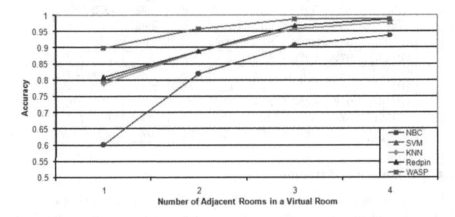

Fig. 3. Accuracy of each virtual room

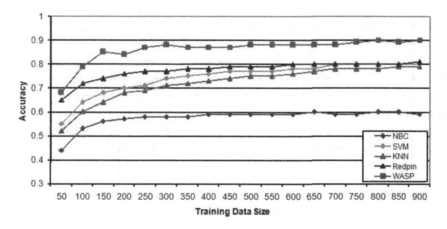

Fig. 4. Accuracy of different training data size

When one adjacent room is added, the accuracy of all algorithms can be enhanced to over 80%. When four adjacent rooms are combined, the accuracy can even be better than 90%. Therefore all algorithms are more accurate when estimating a coarse-grained location.

4.5 Impact of Training Data Size

Fingerprint approach requires labeled data collection in advance. It is not a trivia task to collect hundreds or thousands of data points. To see the precision of the five algorithms for different training size, we run an ablation study by increasing the training data size from 50 to 900 and evaluate the accuracy on the same testing data (Fig. 4). We choose the training data to ensure that each room has enough coverage. When the data size is 50, the accuracies of all algorithms are not very good. When the data size is 150, the WASP algorithm can give over 80% of the total accuracy, which is better than other alternatives. However, when the number of historical fingerprints is increased, the accuracy improvement is less apparent. One plausible reason is that while more fingerprints provide more matching samples, they also provide more polluted data which confuses the algorithms and reducing the accuracy. Since collecting training data with labels requires non-trivial human efforts, this result shows that even a small amount of training data (e.g. 150) can already provide reasonable indoor locationing accuracy.

4.6 Number of APs

When irrelevant APs are removed for a location, we observe that the indoor locationing accuracy improves. Currently NF chooses these APs occurring more than the average frequency as "relevant APs". We want to see how many relevant APs for one room are needed for acceptable accuracy. We sort the APs based on their

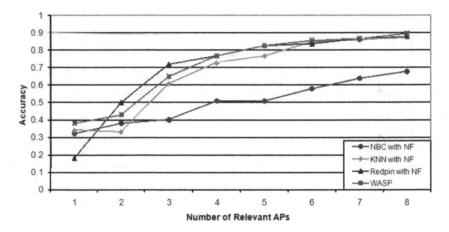

Fig. 5. Locationing accuracy based on different number of relevant APs

relevance for each room. We first choose the most relevant APs and estimate the accuracy by adding the APs one by one. In this experiment we do not include SVM because the probabilities of different training models calculated by LibSVM were not correlated. The accuracies of different numbers of APs for the other four algorithms are shown in Fig. 5. With small numbers of APs (from two to four), the Redpin algorithm has much better accuracy than KNN and WASP. One plausible explanation is that the bonus-penalty adjustment of Redpin makes the discrepancy of two fingerprints more obvious but the PMI cannot provide any additional benefit because the filtered APs are already highly relevant to the locations.

To understand how many APs are needed for reasonable indoor locationing accuracy, we sort APs according their visibility from the highest frequency to the lowest. For each run, we increase the number of APs from 1 to 16 to calculate the

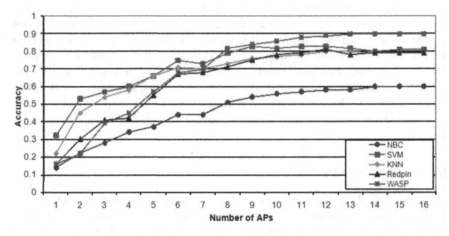

Fig. 6. Accuracy of a different number of APs

accuracy. The result is shown in Fig. 6. We see that SVM has better accuracy when the number of APs is fewer than ten because the primary APs can provide enough information for the SVM classification. However, when the number is more than ten, the SVM classification is affected by fingerprint pollution and the result is worse. On the other hand, the Redpin algorithm performs better than WASP when the number is fewer than three. However, when the number is more than seven, WASP outperforms Redpin by around 10%. Another interesting observation is that the improvement in accuracy slows after nine APs. It seems that these less visible APs do not provide essential information for location detection. Instead, they may even cause confusion when matching fingerprints and lower the accuracy.

5   Conclusion and Future Work

In this paper, we propose WASP, an enhanced indoor locationing algorithm for a congested Wi-Fi environment. Our approach takes signal strengths, AP visibility and statistical fingerprint history into consideration to enhance the Redpin algorithm in a congested Wi-Fi environment. This approach obtains the best accuracy and also works well even with the small training data set in the experiments. Even though WASP may not work well with a small number of APs, most office buildings and homes are covered by more than three APs and the fluctuations and congested signals are likely to be more serious in a real world than in the laboratory. We believe WASP can provide an overall satisfying indoor locationing prediction.

In this paper, we only chose nine public rooms on the second floors. We plan to extend the collection to more private and wall-bounded rooms over two floors. Multiple RF fingerprints[5], such as Bluetooth, might also improve the accuracy. In addition, we will explore the use of accelerometer data to determine if a user is moving or not and thereby enable time-averaging or tracking to improve accuracy. Another interesting issue is to study the optimal number of APs and their positions in the building for the best indoor locationing accuracy. Finally, we plan to apply the WASP algorithm to several mobile health and mobile professional applications. We will also design more incentive and intuitive ways to collect the fingerprints through users' collaboration.

Acknowledgements

This research was supported by grants from Nokia Research Center and by CyLab at Carnegie Mellon under grant DAAD19-02-1-0389 from the Army Research Office. The views and conclusions contained here are those of the authors and should not be interpreted as necessarily representing the official policies or endorsements, either express or implied, of ARO, CMU, or the U.S. Government or any of its agencies. We want to thank the initial assistance of Philip Bolliger

---

[5] The original Redpin used Bluetooth and GSM cell-ID to augment Wi-Fi.

to get us started using Redpin. We also acknowledge the efforts of Joshua Correa and Ed Katz, who explored an earlier set of approaches to Wi-Fi based location in our building, to Patricia Collins who reviewed multiple versions of the paper.

References

1. Bahl, P., Padmanabhan, V.: Radar: an in-building rf-based user location and tracking system. In: IEEE INFOCOM 2000, vol. 2, pp. 775–784 (2000)
2. Barcelo, F., Evennou, F., de Nardis, L., Tome, P.: Advances in indoor location. In: LIAISON - ISHTAR Workshop (Septemper 2006)
3. Bolliger, P.: Redpin - adaptive, zero-configuration indoor localization through user collaboration. In: ACM International Workshop, March 2008, pp. 55–60 (2008)
4. Brunato, M., Battiti, R.: Statistical learning theory for location fingerprinting in wireless lans. Computer Networks and ISDN Systems 47, 825–845 (2005)
5. Carlotto, A., Parodi, M., Bonamico, C., Lavagetto, F., Valla, M.: Proximity classification for mobile devices using wi-fi environment similarity. In: ACM International Workshop, pp. 43–48 (2008)
6. Correa, J., Katz, E., Collins, P., Griss, M.: Room-level wi-fi location tracking. CyLab Mobility Research Center technical report MRC-TR-2008-02 (November 2008)
7. Fan, R., Chen, P., Lin, C.: Working set selection using second order information for training support vector machines. Journal of Machine Learning Research 6, 1889–1918 (2005)
8. Hightower, J., Borriello, G.: Location systems for ubiquitous computing. IEEE Computer 34, 57–66 (2001)
9. Ho, W., Smailagic, A., Siewiorek, D., Faloutsos, C.: An adaptive two-phase approach to wifi location sensing. IEEE International Conference 5, 456 (2006)
10. Li, B., Salter, J., Dempster, A., Rizos, C.: Indoor positioning techniques based on wireless lan. In: IEEE International Conference, p. 113 (March 2006)
11. Paschalidis, I., Lai, W., Ray, S.: Statistical Location Detection. In: Mao, G., Fidan, B. (eds.) Localization Algorithms and Strategies for Wireless Sensor Networks: Monitoring and Surveillance Techniques for Target Tracking, IGI Global (2009)
12. Seshadri, V., Zaruba, G., Huber, M.: A bayesian sampling approach to in-door localization of wireless devices using received signal strength indication. In: IEEE International Conference, March 2005, pp. 75–84 (2005)
13. Wu, C., Fu, L., Lian, F.: Wlan location determination in e-home via support vector classification. In: IEEE International Conference, vol. 2, pp. 1026–1031 (2004)

A Long-Duration Study of User-Trained 802.11 Localization

Andrew Barry, Benjamin Fisher, and Mark L. Chang

F.W. Olin College of Engineering, Needham, MA 02492
{andy,benjamin.fisher}@students.olin.edu,
mark.chang@olin.edu

Abstract. We present an indoor wireless localization system that is capable of room-level localization based solely on 802.11 network signal strengths and user-supplied training data. Our system naturally gathers dense data in places that users frequent while ignoring unvisited areas. By utilizing users, we create a comprehensive localization system that requires little off-line operation and no access to private locations to train. We have operated the system for over a year with more than 200 users working on a variety of laptops. To encourage use, we have implemented a live map that shows user locations in real-time, allowing for quick and easy friend-finding and lost-laptop recovery abilities. Through the system's life we have collected over 8,700 training points and performed over 1,000,000 localizations. We find that the system can localize to within 10 meters in 94% of cases.

1 Introduction

Computerized localization, the automatic determination of position, will augment existing applications and provide opportunities for new growth. One can easily imagine a phone, computer, or other device changing behavior based on location. A phone might disable its ringer when in a conference or classroom. Calendar reminders would only appear if a user was not already in the event's location. A laptop could automatically select the closest printer when printing. Finding a colleague would be as simple as looking up a phone number.

Localization abilities have spawned a number of companies including GPS navigation [5][16], asset tracking [4][17][20], and E911 systems [6]. The most common form, GPS, performs well in many instances, but it cannot achieve good accuracy indoors. To provide indoor localization, researchers have examined the use of dedicated hardware including ultrasound, IR, and RF beacons. Most of these platforms provide good resolution, but often have high installation, maintenance, and usage costs. With the advent of 802.11 wireless networking, researchers have turned to utilizing wireless access points as fixed RF beacons. This method mitigates the high hardware and installation costs of earlier systems, but often requires a substantial amount of off-line training, or the collection signal strength samples in many locations. Here we describe a large scale deployment of a system that uses 802.11 access points to localize, but transfers the training burden to the system's users, providing a cheap, fast, accurate, and low maintenance method for automated indoor localization.

R. Fuller and X.D. Koutsoukos (Eds.): MELT 2009, LNCS 5801, pp. 197–212, 2009.
© Springer-Verlag Berlin Heidelberg 2009

We use the following terminology to describe our system: a wireless *fingerprint* denotes the signal strengths of surrounding access points at a given location. A *bind* is the act of associating a fingerprint with a location. An *update* is a location scan and localization calculation. Fingerprints are collected automatically, binds are performed by users, and updates are performed automatically or by user request.

2   Related Work

Location-aware computing is not new. Perhaps the best-known location-discovery platform is GPS, which uses U.S. government satellites to compute latitude and longitude [8]. While the high costs associated with GPS systems have disappeared, a receiver can only obtain a location with a clear sky-view. Moreover, GPS experiences substantial drift and is often not accurate enough to obtain room-level localization. Another set of commonly available localization systems serve the FCC's E911 initiatives [6]. These systems focus on approximately 100 meter accuracy and thus, like GPS, are of limited utility in indoor, room-level environments.

The first indoor location-aware systems, such as Active Badge and MIT's Cricket, succeed with specialized hardware [13][18]. Active Badge uses wearable transmitters and a network of sensors to gather location information and report it back to a server. The Cricket system uses a combination of RF and ultrasound to provide accurate and private location data. These systems avoid training, but instead require a substantial hardware installation phase. Both Active Badge and Cricket require location-bound hardware that necessitates prior access by trained personnel to each desired localization area. This installation and the associated time and hardware costs limit these systems' wide-scale use.

As 802.11 networks became common, researchers began utilizing existing hardware to compute location. Microsoft's RADAR and later Haeberlen *et al.* show success in using the signal strength of 802.11 nodes to determine fine-grained indoor location [1][7]. These and similar systems [11][14] require specialized training to create a database of location–signal strength tuples. Training demands a substantial upfront effort and physical access to all of the desired areas. Moreover, after some time, the training data needs to be refreshed to account for changes in the environment and access point locations.

To reduce the expense of training, Intel Research demonstrates an algorithm that can estimate location with only minimal data by expanding its known area with continued use [10]. While the self-mapping algorithm costs little to implement, it requires a significant period of time to gain acceptable localization accuracy and coverage. Moreover, multiple radio configurations complicate the implementation of a shared-training system. Wardriving can be used to seed the algorithm, but those data are often not dense enough for accurate indoor localization.

Both Bolliger and Teller *et al.* introduce crowdsourcing methods that allow users to train and correct the system [2][15]. Teller's work, conducted in parallel with our own, is similar to the system described here although on a smaller scale in time, space, and number of users. They studied 16 trained users limited to a single building with a specialized platform for only 20 days. Here we present a year-long deployment of a similar crowdsourcing method with over 200 untrained users spanning five buildings operating on personal laptops. Bolliger's system engineering is also similar to our own, but, again, he does not present results from a significant deployment.

In the commercial space, Ubisense has deployed accurate localization based on UWB signals in industrial environments [17]. Ekahau has been working on 802.11 localization for a number of years [4]. Skyhook Wireless has combined crowdsourced data with a substantial set of training data to improve their worldwide 802.11 localization system [14]. Navizon uses exclusively user-produced data, but like Skyhook, their system focuses on outdoor localization [12].

3 Architecture

3.1 Overview

Like [2] and [15], we use a client-server architecture to enable fast, accurate localization and provide a mechanism for feedback. To localize, clients perform an *update* in which they collect wireless 802.11 signal strength information (a *fingerprint*) to send to a server. The server computes the client's location and sends the estimate back to the client for optional user review. The server also updates the friend-finding interface with the client's new location in case another user wants to locate the first. When the client receives the location estimate, it offers the user an opportunity to confirm or correct it (Figure 1). If the user chooses to take this opportunity (*binding* a fingerprint), the client sends the new ground-truth data back to the server, which stores the record for use in all future localization computations.

Our system architecture is similar to MIT's Organic Indoor Location system (OIL) [15] with a server-based localizer and without client-side caching. For brevity, we will examine only the novel aspects of our system in depth and direct the reader to the OIL implementation for other details. To compute locations, we use a Euclidean distance algorithm, comparable to RADAR's Nearest Neighbor in Signal Space (NNSS) [1]. We have implemented the algorithm in SQL, interfacing PHP and wxPython clients that run on Windows, Linux, and Mac operating systems. We are planning to implement clients for smartphones and PDAs in the near future.

3.2 Deployment Site

A number of our design decisions were driven by the context in which our system was deployed. We developed and continue to run our localizer at Olin College, a small residential engineering school near Boston. Olin houses its entire 300-student population on campus with five buildings in total, encompassing more than 300,000 square feet. Our primary user base is students, with faculty and staff comprising less than 3% of users. Each student owns an institution-issued laptop although some use their own systems. Since each entering class has a slightly newer laptop model, we find a variety of similar but distinct radio/antenna combinations on campus.

In asset tracking, medical, or warehouse situations, one might choose to localize every few seconds, but students remain in the same place, be it a classroom, library, or residence hall, for extended periods of time. Thus, we chose to localize once every five minutes, an unscientifically chosen but reasonable interval, in order to preserve system resources. Finally, we note that with the system providing a useful service, users have an incentive to provide accurate data and very little impetus to falsify locations.

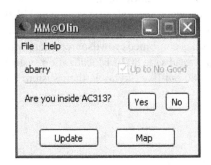

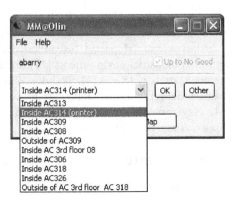

Fig. 1. Typical interface. The client has localized and asks the user to confirm its estimate. The upper left displays a username and the upper right contains a humorous checkbox.

Fig. 2. Training interface. The user has clicked "No" in Figure 1 and is now prompted with nearby locations. Clicking "Other" allows the user to create a new location point.

3.3 Client Interface

The user interface is designed to be as non-invasive as possible. Since the system runs on personal laptops, it must have a small resource footprint minimizing CPU, memory, and power consumption. The client autostarts minimized in the system task tray and, to encourage use, never prompts the user without request, even when the location estimate is known to be poor. We found that we could collect enough training data without interrupting users and annoyed far fewer people in the process. If users want to train the system or access others' locations, double-clicking the task-tray icon brings up our deliberately simplistic interface (Figure 1). When an update is performed, the client displays its location estimation in question form, prompting a training response. The user can then accept the given location, choose from a list of likely locations, or create a new point. Figures 1–3 show typical GUI screens.

Should the user determine that the localization is not correct, he or she can create a new location point. The client prompts for the building, floor, and a text name,

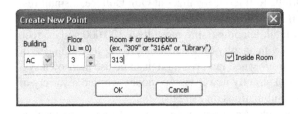

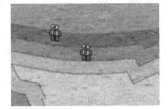

Fig. 3. New location creation interface. The user has clicked "Other" in Figure 2 and is now prompted to enter the details of his or her location. After clicking "OK," the user will be prompted to select the location on a map of the local area.

Fig. 4. Floor interface. Darker edges indicate a lower floor. In this case, two people are located on the ground floor and two more are located on the first floor.

suggesting a room number or descriptive phrase for the text entry (Figure 3). Once a user has entered those data, a labeled map appears allowing the user to select where the new location will appear visually. We found that making this process intuitive is key to ensuring the success of a crowdsourced data collection application.

New point creation is challenging for both the user and GUI designer. Most users are unable to correctly identify their location on an unlabeled blueprint, so the user interface must be copiously labeled to prevent errors. We chose to allow users to customize location names, making their descriptions much more useful to others. For example, instead of calling a room "335," users labeled it "3rd floor lounge."

Allowing for free-form input, however, provided substantial possibilities for error. Although "lounge" is a natural input, it is not useful without context. To obtain both flexibility and accuracy, we constrain building and floor choices while allowing for a textual description. Thus, points have reliable context information and custom labels.

Finally, we note that custom naming lowers the entry barrier for new localization systems. Where we need only rough building diagrams, other systems require fully digitized blueprints or CAD models to generate an initial map.

3.4 Friend-Finding Service

To motivate users to train the system, we implemented a friend-finding service that publishes user locations on a map-like interface. The goals of the service are threefold: allow users to quickly and easily locate specific people, display all users' locations on one screen, and act as an advertisement for the project.

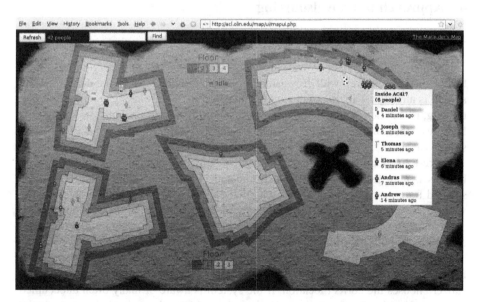

Fig. 5. Friend-finding webpage frontend. The interface is themed like an old magical map. The user has moved the mouse over a cluster of people, prompting a drop-down list of location, names, and update times.

To satisfy our goals, the interface must allow for both searching and browsing while maintaining an attractive look and feel (Figure 5). To encourage use and help explain the concept, we themed the project as the "Marauder's Map at Olin," a reference to a magic map that displays people's locations in the popular Harry Potter series. While this theming might seem trivial, we found that it was critical to building and maintaining a user base. The map theme helped people understand why the service is useful and encouraged them to share with their friends.

Displaying hundreds of users on multiple floors proved to be a challenge. Early tests showed that users clustered near the edges of buildings, so we expanded our building representations to show the edges repeatedly, utilizing the new space to indicate vertical displacement and avoid icon overlap (Figure 4). This is not a perfect solution, as some buildings have popular locations in their centers, but we found these overlaps to be relatively minor.

3.5 Privacy

Privacy is a concern in any localization system. Given that the primary application of our implementation is a friend-finding service, we found that concerned users were aware of the implications and simply chose not to participate. Some users requested the ability to remove their location report at any time, a feature we implemented. All of the system's services reside only on internal servers to ensure that location information is not published outside of our institution.

4   Approach to Crowdsourcing

4.1   Motivation

The primary motivation for crowdsourced data is the reduction in time required to train the localizer. Moreover, with crowdsourcing, users provide most of the data while the system is already localizing, reducing the time before the localizer can be used. When training, researchers found that gaining access to private and semi-private spaces (offices, residence hall rooms, etc.) was difficult and awkward, a problem that user-trained systems avoid [7].

A pre-trained system requires retraining to account for changes in the environment, but a user-trained system is continuously updated with no overhead. In addition, crowdsourced training naturally produces data that are dense in places that are commonly visited. Users tend to bind in places they frequent, causing common locations to have dense data. Because our localizer treats each bind separately, it weights common locations more heavily, resulting in a natural location-frequency dependency.

Finally, traditionally trained systems suffer from a conflict between coverage, the number of distinct locations with data, and accuracy, the measure of how often the localizer chooses the correct location. If the system is aware of many often unoccupied locations, it will suffer from a decrease in accuracy. Crowdsourcing helps mitigate the problem by naturally ignoring rooms that users do not frequent. For example, there are small, narrow trash rooms in each wing of the residence halls that were never bound in our system (Figure 6). A traditionally trained system might incorrectly place a user in

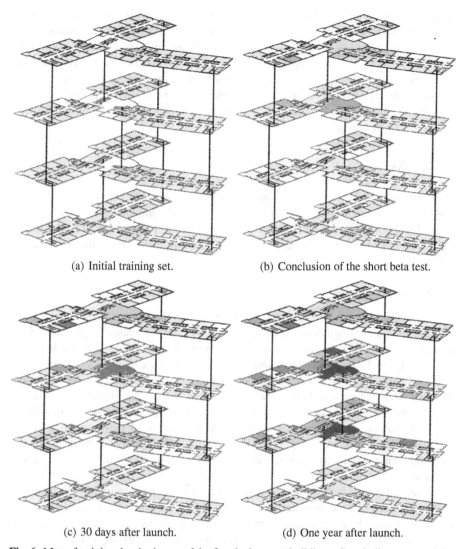

(a) Initial training set.   (b) Conclusion of the short beta test.

(c) 30 days after launch.   (d) One year after launch.

Fig. 6. Map of training density in one of the five deployment buildings. Gray indicates no training data. Light to dark red indicates progressively more fingerprints for that space. Note that common areas (center of the building) have far more data than individual's rooms.

one of these rooms, even though the probability of a user being located there is very small. Our system will always avoid these areas because no users have bothered to train them.

4.2   Initial Training

While almost all of our data are provided by users, we found that a minimally trained system was important to convince users that the software was compelling. Typically, this type of initial training takes about 1-3 minutes per location [7]. At this rate, a system covering over 350 spaces would take approximately 16 person-hours to train manually, but we can train the system to a minimally usable state, in about 1.5 hours. With that training, we create a sparse map including hallways and common areas, allowing for reasonable (within 10-20 meters) estimates that support further training by users. This training is easy to perform because all fingerprints can be collected in public areas and with relative infrequency. To train our system we simply walked down most hallways and bound one or two points per hall.

5   Results

After a short beta test, we deployed our system campus-wide at Olin College in April 2008 and have continued operations for over a year. We announced the project with an email to a common list in which we explained the concept and encouraged users to download and run the client. To date, we have had more than 200 unique users, 8,700 binds (95% of users contributing), and over 1,000,000 location updates.

5.1   System Accuracy

A localization system's performance is determined by both *coverage* and *accuracy*. Coverage measures how much of the deployment area has associated fingerprints while accuracy measures of how often the localizer reports the correct location. We first discuss coverage and then proceed to examine our localizer's accuracy.

At the beginning of deployment, only our initial survey supplied data, so we started with poor coverage, especially in private rooms (Figure 6(a)). We found that within 30 days of launch, our coverage stabilized at a reasonably complete level, with over 75% of all known locations at the year's end having already been bound. Coverage progression for one building can be see in Figure 6. Other buildings show similar patterns, although places students are unlikely to spend time, such as faculty offices, never achieve good coverage.

Our second metric is system accuracy. Accuracy starts poor and improves with the number of binds. To measure accuracy, we note that we should not simply test the system at random locations. Real accuracy is determined by how often the localizer correctly estimates *users'* locations. To use the aforementioned example, the localizer's performance in a small trash room is of little importance to true accuracy.

To test our system in this manner, we chose to survey our users during deployment. After the system had reached a steady state, we emailed our users asking them to report the localizer's accuracy at that moment. Thus, users opened the client, checked its current estimate against their real location, and reported performance in meter ranges (ie within 0-5 meters, 5-10 meters, etc.). We received 57 reports representing more than half of all online users within 8 hours. While these reports were self-selected based on which users chose to respond, we do not believe this bias has skewed our data measurably.

Figure 7 shows localization errors. We find that we localize to within 5 meters in 69.9% of attempts and to within 10 meters in 94.9% of cases. This accuracy is approximately equal to other published systems, although it does not achieve the performance of Haeberlen *et al.*'s calibrated or King *et al.*'s dense data techniques [7][9][15].

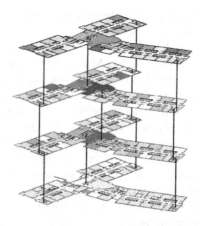

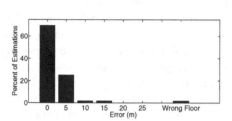

Fig. 7. Localization error. We determined error by asking users to perform spot checks in their current location. We find that we localize correctly in 69.9% of attempts and are within 10 meters in 94.9% of cases. We localize to to the wrong floor only 1.8% of the time.

Fig. 8. Combined density of location reports for one year in one residence hall. Clearly, common areas are visited far more often than individuals' rooms. No students live on the first floor so we are not surprised to see few reports in that location.

5.2 System Vulnerabilities

We note a number of conditions that degrade our performance. First, during our deployment, a large fraction of the campus's access points received firmware modifications that resulted in a change in MAC address. The system does not recognize the new configurations and assumes that none of the old access points exist. While our architecture is designed to easily adapt to new access points with a single dictionary replacement, we found that losing these access points did not significantly degrade performance, so we allowed the system to operate without intervention.

In addition to changing MAC addresses, network administrators moved a small fraction of access points to improve wireless performance. While users have retrained the system, this movement continues to degrade our performance. To mitigate this issue, we are considering a weighting system that favors new fingerprints, marginalizing old and possibly outdated data. The design of the weighting system requires further study to determine if it should universally downgrade old data or only ignore old fingerprints when newer ones are available for a location. The first implementation would cause the localizer to favor newly bound points while ignoring old fingerprints, effectively reducing coverage over time. The second implementation retains coverage, but might not reflect newer user-movement patterns.

A third potential degradation of performance is the automatic and manual gain control on access points. During our deployment both automatic and manual power adjustments were made, but the system showed no noticeable decrease in performance. Without a clear indication that this was causing degradation of localizations, we have not spent the engineering resources to study this effect further. It is worth noting, however, that these power changes affect only the local area around the modified access points and simply shift the localizer's tendency closer or further from the respective wireless node.

A fourth potential degradation of performance is our support of a wide range of laptops, including Mac hardware and at least four models of the Dell Latitude D-series. We do not currently account for radio/antenna configurations, although the system has a natural correction based on user binding patterns, as described below.

In the user space, we note that user-generated data are not as reliable as professionally generated scans. It is difficult to determine how often localization errors are due to user error when training the system or from signal strength variations resulting from dynamic objects, antenna orientation, radio chipset, access point power fluctuations, or a host of other phenomenon. General accuracy statistics provide some insight for an upper bound on errors, but we have not yet developed a metric to study these errors explicitly. A time-based weighting on data, as examined above, would provide some mitigation for mistaken inputs, allowing them to be corrected as new, better data become available.

Finally, our system does not address the possibility of malicious users, beyond marking each bind with an identifier that allows the elimination of all binds by a particular user. While this is a concern, we are not aware of a single instance of such activity. Moreover, to be effective, malicious users would need to create a substantial number of false data points in many locations to overwhelm the existing fingerprint set. Many binds in one location would overwhelm that particular location, but the damage would be confined to the local area. To be truly successful, a malicious user would need to bind incorrect data throughout the system's coverage, a far more difficult task.

In the event that malicious activity becomes an issue, we have discussed the implementation of weights based on how similar new data are to existing fingerprints. In this way, outliers are rendered harmless automatically. If a significant number of outliers were added to the system by different users, the weights would begin to skew towards those new fingerprints, accounting for dramatic environment changes such as access point relocation.

5.3   User Behavior

After release, we collected about 71 binds per day and within 2 months our data set had grown to 27 times the initial training, a collective effort of approximately 25 person-hours in ideal conditions. While training rates decreased as accuracy increased, users still train the system over one year later. Figure 9 shows these trends in database size over time.

As expected, we collected more fingerprints in common and public locations than in private rooms. The median number of binds per room is 7 and the maximum is 305, which occurs in a residence hall lounge (a common socializing space with couches and a TV.) 17% of known locations have only one bind and 53% have 10 or fewer. As the system's coverage grew, the number of new locations bound quickly decreased. Our initial survey bound 16% of total locations, beta testers bound 48%, and our general user base bound the remaining 36%. Thus, we find that users are far more likely to pick an existing location than they are to bind a new one.

In addition, we find that the user contribution profile is similar to other mass-interaction applications [15][19]. A few enthusiastic users bind an inordinate number of times. Interestingly, these users do not update their location as often as we might expect, with very little correlation between bind and update frequencies.

We hypothesized that our data collection method would result in dense data in frequented places, and our expectations are confirmed. We see this in the similarity between Figures 8 and 6(d) which show where locations are reported and bound, respectively. These data confirm our second hypothesis, that users bind in places that they, individually, frequent. 51.2% of all location updates occur in places that users had bound themselves. In other words, half the time a user is in a place where he or she has contributed data. Thus, we confirm that one of the primary advantages of crowdsourced data collection is that users are willing to train where they frequent and that they tend to reside places where their own data are best.

5.4 Application Use

While we have not yet performed a formal study, we feel that users are satisfied with the localization application. In one year we logged over 14,000 friend-finding page loads, averaging well above 50 hits per day during both fall and spring terms. We note that not only did users utilize the system at launch, they continue to use the service throughout its deployment. We also note that with 100 active users, we are localizing about 1/3 of the student population's systems.

Our last method of evaluating user satisfaction is purely anecdotal. Users tell us that they enjoy using the system and have only rarely contacted us with complaints, despite our contact information being readily available. Perhaps our favorite example of users' creativity is using the system in a scavenger hunt. An on-campus business group was running a promotion in which they planted clues advertising the whereabouts of free product samples. To create one of the clues, the group, unbeknown to us, managed to emulate a client and change the reported name and user icon to their name and logo, respectively. They then used the reported location to advertise where free samples could be found. Months later, when performing data analysis, a developer found the odd entry and finally traced it to the group who reported that it was the most popular clue in their entire game.

6 Detailed Analysis

We now discuss the system's usage in detail. We examine trends in both when and which users train the system and the profile of where users localize. We find that new users are the primary providers of ground-truth data and that, despite our high participation rate, a small set of users provide most of the system's data. These often-training users, however, are neither more or less likely to localize than their non-binding peers.

6.1 Training Rates

Users tend to bind data in their first few days of use and rapidly stop providing ground-truth information thereafter. Throughout the system's life, 43% of all binds occur within

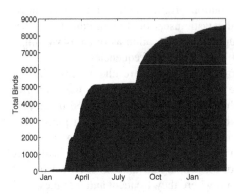

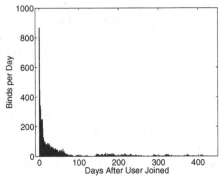

Fig. 9. Fingerprint database size over time. Training rates were steady after release and have reduced as the system became more accurate. We are not surprised to see little data added during the summer term when students are not on campus.

Fig. 10. All binds sorted by days since the binding user first localized. Clearly, users bind a significant amount in their first day and steadily less throughout their usage. We find that 11.5% of all binds occur on a user's first day.

10 days of the binding user's first application use. We offer two explanations for this phenomenon. First, users may train the system because of the novelty of providing data. Once that novelty fades, users become less interested and only localize. Second, student movement patterns do not change substantially from one day to the next. Once

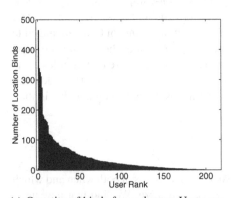

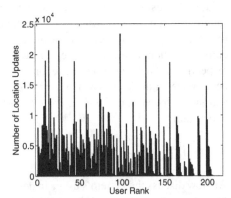

(a) Quantity of binds for each user. Users are sorted and assigned an ID by decreasing number of binds.

(b) Number of updates per user. Users have the same ID as in (a).

Fig. 11. Number of binds and updates for each user. Users are sorted by decreasing number of binds. In (a) we see that 20% of users bind 66% of the ground-truth data. In (b) we find that those often-binding users are not significantly more likely to update their location often, showing that the primary producers of data are not necessarily the primary consumers.

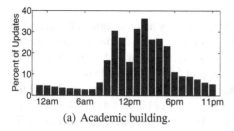

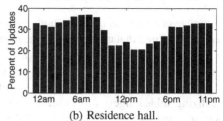

(a) Academic building.　　　　　　　(b) Residence hall.

Fig. 12. Percentage of localizations per hour in both an academic building and a residence hall. As expected, we find that users occupy academic buildings during daylight hours (excluding lunch) and residence halls at night. We also note that while users arrive to classes in large groups, they tend to leave in a more gradual manner.

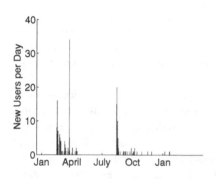

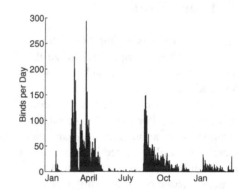

Fig. 13. Number of new users per day. Our campus-wide release occurred in April and a new academic year started in September.

Fig. 14. Binds per day. Given that new users bind data often, we are not surprised to see a significant correlation to Figure 13. One interesting case is the system's second January, where we see no new users but a significant number of binds. In this period, students return to campus after intersession and appear again interested in binding data.

a user has trained his or her habitual places, the system may not require further training for accurate use and thus exclusively localizing is acceptable. Figure 10 shows bind occurrences sorted by days since adoption.

While new users train the system more often than longtime users, we find that some provide a disproportionate amount of data. Figure 11(a) shows a sorted profile of users based on number of binds. In our system, 20% of users provide 66% of the data. As mentioned above, we notice that there is little correlation between users who provide data and those who localize often. This is manifested in the difference between 11(a) and 11(b).

6.2 Application Usage Patterns

We find cyclic patterns in usage, both throughout each day and throughout the year. As expected, we find that users cluster in academic buildings during the day while returning to residence halls at night. Figure 12(a) shows localizations on an hourly basis in an academic building while 12(b) displays a similar plot for a residence hall.

We find that long-term training patterns are driven by new users. It appears that while localizing holds longevity as a useful service, the novelty of binding data fades, causing users to stop binding. This trend might also be influenced by an increase in accuracy in the places that users individually frequent, requiring less training as the system learns from the user. In addition to Figure 10, we see this pattern in the similarity between Figures 13 and 14, displaying when users join and when users bind, respectively.

7 Future Work

This paper presents an implementation of a user-trained localization service covering an entire college campus. To utilize this framework in other scenarios, we consider a number of additions including new applications and novel ways to collect training fingerprints.

To augment our user base, we are considering implementing additional localization-based services. For example, we might create a tagging system for files that records location information on creation and modification. With this information, a user could search for all files created in a particular room. Users tend to create different types of files in different places, such as minutes in a meeting room and source code in a lab, so searching based on location might be helpful. Other potential services include location-based printer selection and an API to support new developers and new deployments.

Porting our code to hand-held and smart-phone devices would provide interesting new data sources and challenges. Compared to laptops, these platforms are more ubiquitous and would provide more continuous and varied data with their constantly active radios. This more diverse set of hardware might require a platform identification mechanism and conversion functions between device fingerprints, like Haeberlen's implementation across different radio platforms [7].

To gather more binds, we are considering automatic calendar integration. Many calendar appointments are tagged with a location which we could extract and utilize to automatically train the system. When a user's calendar indicates that he or she is in a specific place, the system could automatically collect fingerprints and bind them to that location. Obviously, users and/or their wireless devices are not always located where their calendar indicates, but with intelligent use of idle-times, a limited fingerprint set, and perhaps a movement detector like [3], this type of training could be made reliable. In more extreme cases, we might consider using calendar integration to train an entire system without any user interaction, although the accuracy of these fingerprints would require careful study.

Finally, we are continuing analysis of our data and are planning more formal user surveys to better characterize the system's strengths and weaknesses. Moreover, we are planning more expansive accuracy experiments that will inform the distinction between random-location accuracy and the common-area accuracy that a normal user experiences.

8 Conclusion

We have described a long-running test of a user-trained system that performs accurate indoor wireless localization in areas with existing 802.11 networks. The system can be deployed in any location that has a pervasive network and a group of users willing to train it. By utilizing personal laptops and existing access points, we do not need to build or buy any additional hardware. The system's interfaces are simple and intuitive, allowing users to localize, find others, and contribute training data with no instruction.

After more than one year, our system continues to operate and has accumulated more than 200 users, 8,700 binds, and over 1,000,000 location updates. Usage patterns provide natural guidance for the localizer, improving accuracy by accumulating dense data in common areas. These methods result in successful localization to within ten meters in over 94% of cases, providing convincing evidence that crowdsourcing is a practical method for cheap, pervasive wireless localization.

Acknowledgements

We would like to thank the anonymous reviewers for their valuable comments and guidance. We also thank the community at Olin College for their ideas, testing, and feedback. Andrew Barry and Benjamin Fisher are supported by F. W. Olin Scholarships.

References

1. Bahl, P., Padmanabhan, V.: RADAR: An in-building RF-based user location and tracking system. In: Proc. IEEE Infocom 2000, pp. 775–784. IEEE Computer Society Press, Los Alamitos (2000)
2. Bolliger, P.: Redpin - adaptive, zero-configuration indoor localization through user collaboration. In: MELT 2008: Proceedings of the first ACM international workshop on Mobile entity localization and tracking in GPS-less environments, pp. 55–60. ACM, New York (2008)
3. Bolliger, P., Partridge, K., Chu, M., Langheinrich, M.: Improving location fingerprinting through motion detection and asynchronous interval labeling. In: Choudhury, T., Quigley, A.J., Strang, T., Suginuma, K. (eds.) LoCA 2009. LNCS, vol. 5561, pp. 37–51. Springer, Heidelberg (2009)
4. Ekahau, Inc., http://www.ekahau.com
5. Garmin Ltd, http://www.garmin.com
6. Geer, D.: The E911 dilemma. Wireless Business and Technology (November 2001)
7. Haeberlen, F., et al.: Practical robust localization over large-scale 802.11 wireless networks. In: Proceedings of the Tenth ACM International Conference on Mobile Computing and Networking (MOBICOM) (September 2002)
8. Hightower, J., Borriello, G.: Location systems for ubiquitous computing. Computer 34(8), 57–66 (2001)

9. King, T., Kopf, S., Haenselmann, T., Lubberger, C., Effelsberg, W.: Compass: A probabilistic indoor positioning system based on 802.11 and digital compasses. In: Proceedings of the First ACM International Workshop on Wireless Network Testbeds, Experimental evaluation and CHaracterization (WiNTECH), Los Angeles, CA, USA (September 2006)
10. LaMarca, A., et al.: Self-mapping in 802.11 location systems. In: Beigl, M., Intille, S.S., Rekimoto, J., Tokuda, H. (eds.) UbiComp 2005. LNCS, vol. 3660, pp. 87–104. Springer, Heidelberg (2005)
11. Letchner, Fox, LaMarca: Large-scale localization from wireless signal strength. In: Proceedings of the National Conference on Artificial Intelligence, AAAI (2005)
12. Navizon: Peer-to-peer wireless positioning, http://www.navizon.com
13. Priyantha, N., Chakraborty, A., Balakrishnan, H.: The cricket location-support system. In: 6th Ann. Intl. Conf. Mobile Computing and Networking (Mobi-com 2000), pp. 32–43. ACM Press, New York (2000)
14. Skyhook Wireless, http://www.skyhookwireless.com
15. Teller, S., et al.: Organic indoor location discovery. Technical Report MIT-CSAIL-TR-2008-075, MIT (December 2008)
16. TomTom NV, http://www.tomtom.com
17. Ubisense, Ltd., http://www.ubisense.net
18. Want, R., Hopper, A., Falcao, V., Gibbons, J.: The active badge location system. ACM Transactions on Information Systems 40(1), 91–102 (1992)
19. Whittaker, S., Terveen, L.G., Hill, W.C., Cherny, L.: The dynamics of mass interaction. In: Conference on Computer-Supported Cooperative Work (CSCW) (November 1998)
20. Wisetrack, brand of TVL, Inc., http://www.wisetrack.com

Tutorial on Location Determination by RF Means

Richard Fuller

Location-Based Services Special Interest Group (LBS SIG)
rfuller@lbssig.org

Abstract. This paper focuses on radio-frequency (RF) location determination characteristics and implementations. A presentation of RF transmission, propagation and reception characteristics is provided and a summary of some the major developments of RF-based location systems is also discussed. RF determination capabilities are typed and classified and outlined. Finally, examples of RF-based location systems are given.

1 Introduction

Location determination has a long a storied history. From the first navigators guided by the stars and Sun to the electronic systems of today; finding one's place continues to be amongst the most crucial elements in life. The term "location awareness" has sprung into the popular vernacular in recent years in response to the desire for applications and services to provide greater availability and accuracy of location determination at all times. Location awareness helps provide context for feature rich applications and services; making them feel far more personal and interactive. While the relative importance of location awareness may not have increased in the past few decades, the proliferation of location determination technology for assets and persons has increased remarkably. Many of the location determination technologies we take for granted today such as GPS (Global Positioning System), and its international Global Navigation Satellite System, or GNSS, counterparts such as GLONASS, Galileo, Compass/Beidou, etc., were not available 20 years ago. The introduction of global navigation systems has produced a paradigm shift in location awareness that expects high-availability, always-on location determination capabilities. This expectation has, in turn, led to substantial innovation and development in the location algorithms and technology that today provide a myriad of options for various applications.

This paper provides a tutorial on Radio Frequency (RF) of location determination technology. It starts by looking at RF system characteristics versus other options, such as inertial and vision systems. The review of RF location determination systems will begins with RF bands and their properties. It briefly discusses the impact of bandwidth on location determination accuracy. The paper then discusses the different RF location approaches followed by methods employed in RF location determination.

1.1 Frequency Bands and Usable RF Spectrum

When considering the alternatives for location determination systems, one primary question that must be answered is: "Why use RF systems versus other non-RF

R. Fuller and X.D. Koutsoukos (Eds.): MELT 2009, LNCS 5801, pp. 213–234, 2009.
© Springer-Verlag Berlin Heidelberg 2009

approaches?" First, RF represents some of the most flexible kinds of equipment that can be deployed versus the alternates. For example, ultrasound-based systems can have a very fine precision but are strictly limited to Line-Of-Sight (LOS) operation and highly subject to environmental noise [1]. Table 1 shows the performance of different non-RF systems deployed today.

In most cases the most limiting factor is range. Since RF propagation can be used for close-in applications as well as those applied over hundreds of kilometers, it is the most flexible method for use in applications. However, in many cases where RF has superior range it does so at the expense of precision, which is why alternative means of precise location determination will always be sought after. One such example in wide use today is the compliment of inertial sensors and GNSS. This combination can offer very precise location capability without the need for installed infrastructure. This is only limited by the precision of the GNSS system it is paired with as well as the availability of those precise signals to calibrate the inertial sensors. This type of combination continues to be of great interest in development of commercial and consumer systems.

Table 1. Comparison of Different Location Technologies [1]

Technology	Pros	Cons
Inertial	-Precise with expensive sensors -Works in a variety of environments (underwater, indoor, etc.)	-Unusable without frequent corrections from external reference (often GNSS-based)
Vision	-High precision -Works where instrumented	-Limited range, LOS -Extensive instrumentation
Ultrasound	-Very inexpensive emitters and sensors -High precision	-Limited range, LOS -Very sensitive to ambient noise
Infrared	-Inexpensive emitters and sensors -High precision	-Limited range, LOS -Very sensitive to ambient noise
Smart/Active Floors [2]	-Works without instrumenting subject -High precision	-Works only where floor (or furniture) is instrumented -Difficult to distinguish individuals (i.e., track a specific person)
GNSS	-Worldwide coverage -Free to use	-Precision 10m unaided -Limited indoor use

Table 2. Classification of radio frequencies [3]

Band	Frequency	Wavelength
Very Low Frequency (VLF)	< 30 kHz	> 10 km
Low Frequency (LF)	30-300 kHz	1-10 km
Medium Frequency (MF)	300 kHz-3 MHz	100 m-1 km
High Frequency (HF)	3-30 MHz	10-100 m
Very High Frequency	30-300 MHz	1-10 m
Ultra High Frequency	300 MHz - 3 GHz	10 cm - 1 m
Super High Frequency (SHF)	3-30 GHz	1 - 10 cm

Radio propagation is classified as electromagnetic waves between the frequencies of 10 kHz and 300 GHz. Table 2 shows the classification of RF bands into groups. The relationship between frequency, f, and wavelength, λ, is given by $\lambda = c/f$, where c is the speed of light in a vacuum (2.99792x108 m/sec). The term "microwave" indicates wavelength between 1 m and 1mm.

One of the most critical aspects of RF location systems is the range in which they can be deployed. Systems that use frequencies at and below HF can transmit over the horizon using a surface wave (follows the surface of the Earth). An issue with using the surface wave is that multipath (more than one version of the same signal arriving at the receiver with different delays or angle-of-arrivals). Multipath affects location systems in three distinct ways: 1) it can produce signal fade due to cancelation of waves at the receiver; 2) it can cause multiple versions of the same signal at the receiver that can alter timing; and finally; 3) for direction-finding systems signals can be received from various directions simultaneously. Other issues in systems employing frequencies at HF and below are antenna size, near-far effects, and interference. Antenna size generally scales with wavelength. The simplest designs of dipole or slot antennas are generally $\lambda/2$ in length with short-dipole or loop antennas are as small as $\lambda/10$ [4]. At this size, simple antennas would be 1 m or longer which may be an issue for some portable location applications. The fields around the antenna can be classified by two different regions, near-field (or the Fresnel zone) and the far-field (or the Fraunhofer zone). In the far-field the energy flow is directed radially (field shape is invariant by distance). As shown in Fig. 1, in the near-field the electric field may be significant and the shape of the field will depend on distance from the antenna.

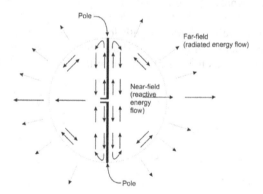

Fig. 1. Near-Field Pattern showing the difference in energy pattern near the element in contrast to the far field which is primarily radial

The near-field radius may be roughly approximated by [4]

$$R = 2L^2/\lambda \quad \text{(m)} \tag{1}$$

where

L = maximum dimension of the antenna, m

λ = wavelength, m

In free-space RF propagation the loss is simply proportional to distance for a given frequency f:

$$\frac{P_r}{P_t} = \frac{1}{(4\pi d f / c)^2} \qquad (2)$$

where Pr is the received power level, and Pt is the transmitted power level. The difference between received power at two distances d1 and d2 is given by:

$$\Delta_P = 10\log_{10}\left(\frac{P_{r2}}{P_{r1}}\right) = 20\log_{10}\left(\frac{d_1}{d_2}\right) \text{ (dB)} \qquad (3)$$

which represents a 6 dB decrease in power for each doubling of distance and a factor of 20 dB with every ten-times increase in distance (i.e., 20 dB/decade). However, due to interference and interaction between the signal and obstacles between the transmitter and receiver along the propagation path, modeling signal loss based only on distance is almost never possible in real-world scenarios. As discussed previously, models of the signal are dramatically impacted by the chosen frequency, the distance between the transmitter and receiver as well as other characteristics. These lead to fading between the transmitter and receiver which have both long-term and short-term components [9]. Long term fading is characterized by terrain variations between the transmitter and receiver, for example is the area flat or hilly. In contrast, short term fading is due to local effects such as reflection and refraction around objects that are on the scale of buildings or a forest. Fading can also be a temporal function based motion of the receiver, or for a stationary receiver, motion of the surrounding objects creating the fading signals. As computed in [5], path loss due to terrain can be an additional 20 dB/decade due to terrain. From foliage, path loss can be roughly 20dB/decade extra as well. Additional path loss can be attributed to buildings and other structures. As reported in [6], and as shown in Fig. 2, structural path loss at 1-2 GHz and 2-4 GHz can be as high as 50 dB for certain structures. Fig. 3 has the mean

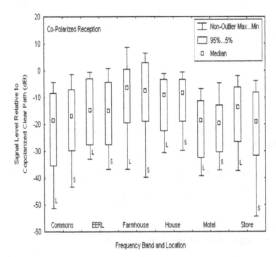

Fig. 2. Attenuation of L-Band and S-Band Signals in a range of indoor settings

attenuation by frequency from 500 MHz to 3 GHz through a farmhouse ranging from 15 dB to 27 dB of loss respectively. It can be seen that frequencies at or above the VHF band, attenuation inside of building can easily exceed 10 dB or more. An advantage to frequencies below VHF is that building attenuation is generally less of an issue. One important note for all frequencies is the so-called near-far problem which is the problem that short distances between a transmitter and receiver can have a huge dynamic range. RF receiver designs usually have a limited dynamic signal bandwidth which is designed for an expected range of received signal powers. This fixes the viable receiver ranges for a given a transmit power and radio-link properties between the transmitter and receiver.

As shown in Table 3, frequencies at MF and below have beneficial signal propagation performance, including good building penetration. However, since efficient antennas at these frequencies are impossible to achieve, high transmit power is required to attain the required signal level at the receiver [4]. For the Loran system which operates in the LF band at 90-110 kHz the transmit power at the stations are 1 MW [1]. While this high power is necessary to operate over long distances (typically 900-1500 km over land, 1500-2000 km over sea [7]) it is also to compensate for antenna losses that could easily exceed 60 dB for a handheld device (<10cm) or 30 dB for a fixed 1 m receiver [4]. While at lower frequencies, efficient antennas are key; at higher frequencies, the local environment becomes the driving factor in location determination. Above 100 MHz, LOS signals become the dominant propagation type but also buildings, vehicles and natural obstructions like trees begin to have greater effect [3]. Multipath becomes a major concern as reflections from these obstacles arrive both in-phase and out of phase at the receiver. Above 3 GHz there is only direct LOS propagation available since obstacles highly attenuate the signal. Multipath is also of substantial concern.

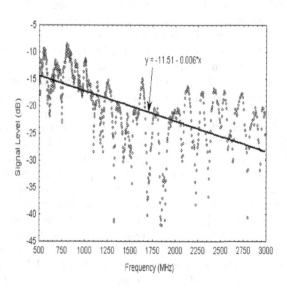

Fig. 3. Dependence on Frequency of attenuation in a farmhouse

Table 3. Characterization of frequency bands

Band	Frequency	Propagation	Antenna Size (half-wave dipole)	Multipath	Structure Penetration
VLF	< 30 kHz	LOS, Surface,Sky	5 km	Ground, Sky	Good
LF	30-300 kHz	LOS, Surface,Sky	0.5-5 km	Ground, Sky	Good
MF	300 kHz-3 MHz	LOS, Surface,Sky	50-500 m	Ground, Sky	Good
HF	3-30 MHz	LOS, Surface	5-50 m	Ground, Artifacts	Poor
VHF	30-300 MHz	LOS	0.5-5 m	Artifacts	Poor
UHF	300 MHz - 3 GHz	LOS	5-50 cm	Artifacts	Very Poor
SHF	3-30 GHz	LOS	0.5-5 cm	Artifacts	Low/None

A great deal of emphasis has been placed in the location determination literature over the past decade using unlicensed-band, short and medium-range communication devices. These devices are generally classified by the terms Bluetooth, ZigBee, 802.11b/g/n (or WiFi), and RFID [1,8]. These devices are generally designed to operate in one or more of the ISM bands (or Industrial, Scientific and Medical as defined by the ITU-R [9]); and must accept any interference from ISM devices (and in turn not interfere with ISM signals). These restrictions make these devices very low power and limited in range. The most popular frequency band in-use today is 2.45 GHz which is used for a broad array of short range communication devices including WiFi, Bluetooth, ZigBee and even some RFID. Also popular is a band at 900 MHz in ITU Region 2 (comprising North and South America, Greenland and eastern pacific islands) and 433 MHz in ITU Region 1 (Europe, Africa, the Middle East west of the Persian Gulf including Iraq, the former Soviet Union and Mongolia). Recently the use of 5.8 GHz has become increasing popular for certain devices. As indicated by Table 3, all of these frequencies perform best under LOS conditions, although some structure penetration (walls and furniture) is assumed, these obstacles can have significant impacts on signal strengths (attenuation) and phase (multipath). Many approaches in the literature either implicitly or explicitly try to take advantage of these propagation characteristics [10]. One problem of the pathloss and multipath issues of these frequencies is that local environmental conditions (placement of furniture, number of people in a room, relative location of receiver/transmitter to a person as well as a host of other variables) can have a strong impact on measurements used in localization. A great deal of effort has been paid recently in the literature on devising algorithms that are more tolerant to these fluctuations [11,12,13]. This will be discussed in greater detail in Section 2.4 of this paper. Another major trend in RF location determination is Ultra-Wide-Band systems or UWB. UWB is defined as a signal with a bandwidth greater than 500 MHz or 20% of the carrier frequency. The benefits of UWB signals for rejection of multipath will be discussed in the next section but the use UWB has its impact. Since UWB signals share the same signal space as licensed and unlicensed bands, they can have an impact on the noise floor for other applications in those bands. In some countries, UWB cannot be used until these issues are addressed [8] and it received limited approval by the FCC in 2002 [33] for data communications, radar and safety applications across a band from 3.1 GHz to 10.6 GHz. The power limits on the FCC order relegates current UWB applications in these bands to indoor and short-range applications. Commercial systems based on UWB are beginning to be introduced [14] and are definitely worth continued consideration.

1.2 Bandwidth Impact on Received Signals

One approach to describe the signal propagation between as transmitter and receiver is using an impulse response [15]. Modeling the signal from the transmitter to receiver separated by a distance, d, as a impulse response, we have:

$$h_d(t) = \sum_{k=0}^{L_p-1} \beta_k^d \delta\left(t - \tau_k^d\right) \tag{4}$$

where L_p is the number of multipath components of the original signal impulse, $\delta\left(t - \tau_k^d\right)$, received with k-th propagation delay τ_k^d. Each of the received multipath signal components have a magnitude and phase, $\beta_k^d = \left|\beta_k^d\right| e^{j\phi_k^d}$, relative to the original signal based on the additive components at the receiver input. Fig. 4 is a representation of the received signal pattern.

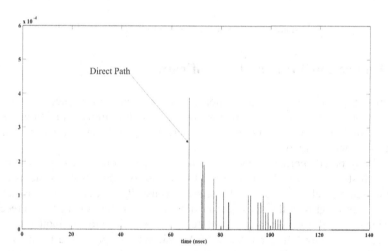

Fig. 4. Multipath profile for an impulse-response channel model

The bandwidth associated with an impulse response is infinite so this channel model has infinite bandwidth. Realizable channels, both because of physical and regulatory limitations, have finite bandwidth (W) and if we represent this signal as $x_W(t)$, the received signal $r_d^W(t)$ is given by the convolution of the limited bandwidth signal and the impulse response [16]:

$$r_d^W(t) = \int\limits_{-\infty}^{+\infty} x_W(\tau) h_d(t-\tau) d\tau \tag{5}$$

Fig. 5 shows a sample channel profile generated by transmitting a raised cosine pulse in a system with 200 MHz of bandwidth. This figure shows how ranging error is observed in the finite bandwidth case versus the unlimited bandwidth case.

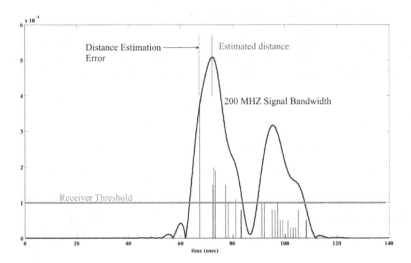

Fig. 5. Distance measurement error due to finite bandwidth signal

2 RF Location Types and Classification

Numerous types of RF location technologies have been implemented over the years. These primarily break down into five major types: 1) Proximity; 2) Direction Finding (DF) or Angle of Arrival (AOA); 3) Doppler; 4) Signal Strength; and 5) Timing or Phase as shown in Fig. 6.

Proximity-based approaches include contact and near-contact sensors such as in RF-ID (which stands for RF IDentification). In the case of RFID, these systems establish location based purely by presence. These systems usually operate in the near-field within a few wavelengths. If the tracked transmitter object is close enough for a receiver to get a signal, then the receiver and transmitter are clearly close to one another.

Direction Finding (DF) and Angle-of-Arrival systems provide a means for determining the bearing of a RF transmitter from a receiver. Two or more receivers can be used to triangulate on the 2D horizontal location of the transmitter (altitude could be determined by a third, vertically-oriented receiver).

Location by Doppler uses the phenomenon that when a transmitter and receiver are travelling towards one another the received frequency is higher than transmitted and when they are travelling away from each other the received frequency is lower. By combining multiple measurements of frequency shifts it is possible to ascertain the location and velocity simultaneously.

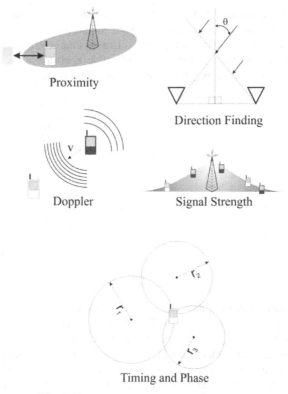

Fig. 6. Types of RF location technologies

Signal strength uses signal power (or other signal-based metrics like Bit Error Rate, or Carrier-to-Noise Ratio) to form an estimate of range or in reference to previously stored location information in a database, otherwise known as "RF finger-printing." In the ranging case, multiple range estimates can be combined via lateration from multiple reference positions. In the case of RF fingerprinting, previously collected reference values are compared to one or more current measurements to match the current location to the stored data.

Timing and Phase location systems use measurements of the received phase of an RF signal or synchronization with a timing modulation on the signal to estimate the range between a transmitter and receiver. Lateration is used to combine multiple range estimates to form a location estimate. This timing information can be employed directly as in Time of Arrival (TOA) systems, or in Round-Trip-Delay (RTD) measurement, or alternatively in Time-Difference of Arrival (TDOA) systems.

The following five sections will discuss in further detail how these different approaches work and look at them in light of various performance characteristics.

2.1 RF Location by Proximity

One of the simplest approaches to RF location is simply by sensing a RF beacon transmitter that is at a known location. In the simplest case, the location precision of the receiver is the coverage area of the beacon transmitter. In other words, the range of the signal represents the uncertainty of your location estimate. Depending on the propagation characteristic of the frequency, channel characteristics, directionality of the transmitter/receiver pair, as well as the effective aperture of the transmitting and receiving antenna, the range may be a few meters to hundreds of kilometers.

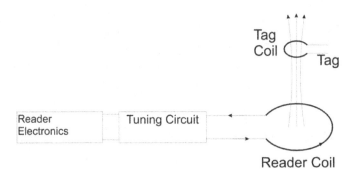

Fig. 7. RFID Operation

The type of device most widely in use today representing proximity-based location is RFID. As depicted in Fig. 7, an RFID system is comprised of at least one "reader" which operates both as a power transmitter and receiver and one or more passive "tags" which are used as transmitters [1]. The system takes advantage of the previously described attributes of near-field RF patterns that allow for electrical and magnetic coupling of the transmitter and receiver. The RFID tags are not usually powered by battery or other power source but derived their power from the reader RF signal. This allows the tag to power a small microchip-based circuit capable of modulating a unique ID code or some other stored information about the tag. Depending on the power transmission of the reader the range to activate the tag can be as little as a centimeter or as large as 30 meters or more [7]. An alternative to passive tags where the reader provides the power for the tag are active RFID tags that have their own power and broadcast information at periodic intervals or when polled by the reader. This approximates a more traditional telemetry system except in this case the tags are designed for minimal complexity, storage and data bandwidth in order to keep their production costs down. RFID operates in different frequency bands. In low-frequency operations it is employed at 30-500kHz and is used mainly for short-range reads (less than 1 meter). Higher frequency system can operate in the range of 850-950 MHz and 2.4-2.5GHz and are designed for longer read ranges.

A novel application that uses a combination of GPS and RFID is the YARD HOUND™ by PINC Solutions [17] which is in commercial use today for management of tractor trailers at distribution centers. This system places passive RFID tags on tractor trailers either permanently or when they enter the yard of a distribution center. A reader is placed on each of the yard trucks that move the trailers between storage bays and the distribution warehouse as well as the exit and entrance to the yard (for ingress and egress records). The yard trucks also have precise GPS location so that when the reader records contact with a particular tag, it can estimate the location of a trailer to roughly one storage bay thus creating a record of the current location of each trailer that they pass. These records are transmitted via 802.11b/g/n wireless link from the yard trucks to a monitoring computer running in the distribution warehouse which hosts a web-based interface that allows both managers inside the warehouse sitting at computer terminals well as workers with mobile devices get instant access to the location of the trailers they need to either fill or empty. This dramatically decreases the time and effort of personnel to survey the yard to map the location of trailers at a given epoch.

Bluetooth (IEEE 802.15) is another wireless technology for short-range communications. It uses a frequency hopping scheme which makes timing/phase measurements very difficult but it is well suited for proximity-based sensing since its range is dependent on its power class (optionally 1, 10 or 100 meters) [18,19]. Bluetooth operates in the ISM band between 2.400 GHz and 2.485 GHz so it has substantial, but not complete, degradation by structures. Each Bluetooth device is identified by a unique Media Access Control (MAC) address. The advantage of Bluetooth is that it is a pervasive technology in consumer and commercial applications. Many of today's cell phones, computers, printers, and other electronic devices come pre-equipped with Bluetooth.

Similar to Bluetooth, WiFi (802.11b/g/n) and ZigBee (802.15.4) RF signals also operate in the ISM band of 2.400-2.485 GHz and can also be used as proximity sensors as well and could be combined with GNSS in a simple either-or scenario for different applications [20]. Further applications for all three of these RF radio systems will be discussed in the sections dealing with Signal Strength for location determination.

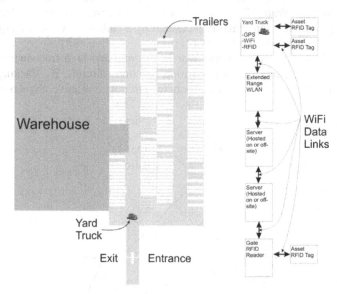

Fig. 8. Elements of the YARD HOUND™ System

In general, proximity-based location can be classified as a ultra-local or local area system where the range generally would not exceed 100 meters. Precision of the location estimate would be inversely proportional to range. Consideration of bandwidth, signal attenuation in the atmosphere or through structures are generally not important because the devices are often operating in the near-field of the RF bands and so far-field effects are inconsequential. Since the relative distance of the receiver and transmitter is so small, antenna gain is usually not a substantial issue so antennas can be small and power requirement generally low. Of particular note, relative to GNSS systems is that these systems always put a constraint on the maximum number of simultaneous users for a fixed infrastructure. Since they all depend on some external reader or basestation which has a limited number of channels these systems will need to increase in size and complexity with an increase in simultaneous users.

2.2 Radio Direction Finding (DF) and Angle of Arrival (AOA)

The basis of radio Direction Finding (DF) and Angle-of-Arrival (AOA) position determination is the use of angular measurements for triangulation as shown in Fig. 9. Triangulation uses the bearings from two or more DF receivers to obtain an estimate of the transmitter's location. Principally these systems operate in the RF far-field of the transmitter and receiver pair with a planar wave incident on the receiver. In practice, multiple signal distortions keep this from being ideal. Work in radio DF dates back over 100 years. The earliest recorded systematic experiments date back to 1899 with the first demonstration of tracking a steamship in 1906 by Marconi using a radial system of horizontal antennas about $\lambda/5$ long. In 1907, work by Pickard and DeForest showed the benefits of loop antenna direction finders which would prove to be the format of choice for many small aperture DF systems (antenna sizes up to about $\lambda/2$). In 1926, the Watson-Watt DF was introduced with used two crossed, quadrature loops that have their output signal directly measured by a cathode-ray oscillograph providing an instantaneous estimate of AOA [21].

In the 1930s leading up to World War II, medium and high frequency radio beacons served guidance and rough navigation functions aircraft. By tracking a single beacon (usually collocated with an airport for aircraft) a relative angular fix can be

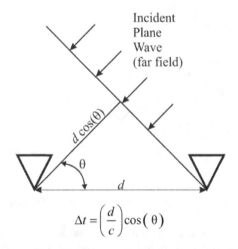

$$\Delta t = \left(\frac{d}{c}\right)\cos(\theta)$$

Fig. 9. DF and AOA Principles

made between the axis of the aircraft and the beacon. By tracking two or more beacons, a crude position could be determined through triangulation. "Positive" navigation points were provided at some airports with vertical marker beacons. These vertical beacons were similar to the standard radio beacon except that their signals were directed upward. Tracking this beacon indicates to the pilot that they are directly above it giving them 'positive' feedback on their location. As more flights were made on a greater number of routes, the use of airport locations as beacons became undesired. Over time, directional beacons were set up along air travel routes around the world. These routes could take advantage of the lessons learned during World War II. During the war, it was shown that routes directed along prevailing winds (referred to as cyclonic) could cut travel time over the shortest distant method (great circle), thus saving fuel. After World War II, VOR (VHF Omnidirectional Range) stations were developed and put into place. According to [22], outside of the developed world VOR support is provided with the exceptions of the Polar Regions, the South Atlantic Ocean, and much of the Pacific and Indian oceans.

VOR broadcasts in a band between 108 and 118 MHz with channels 50 kHz apart (switched to 50kHz from 100kHz in the 1960s to gain additional channels). The ground broadcasts two signals modulated with a 30-Hz tone. The first is a fixed reference tone with an omnidirectional radiation pattern. The second signal is radiated with a cardioid pattern that rotates 30 rotations per second. The receiver uses the relative phase of these two 30-Hz signals to determine bearing to the VOR (which is at a known location). Fig. 10 shows a diagram of a VOR receiver. In addition to the 30-Hz modulation there is an audio channel broadcast modulated at 9660 Hz as well as Morse code identifier of the VOR with a 1020 Hz modulation. The voice signal can be used as a repeating station identifier or as a ground-to-air communications channel [23]. This system design was a U.S. standard by 1946 with later adoption by the International Civil Aviation Organization (ICAO). The performance of this system is only limited by propagation effects and user equipment errors. High-end user equipment can usually achieve 0.1°-10° of angular resolution. This system only works well when the VOR and receiver are line of sight visible [22].

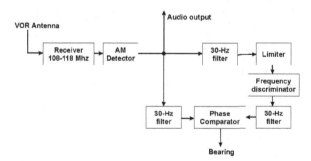

Fig. 10. VOR Receiver Diagram

2.3 Doppler

Another method that can be used to locate a RF device utilizes the well known phenomenon called the Doppler effect. The Doppler effect states that when an object emits a frequency and moves relative to an observer, the frequency of the observed signal will be shifted up or down. The magnitude and sign of the shift depends on the frequency of the signal and the velocity of the transmitter and observer relative to each other. The Doppler shift is given by:

$$\Delta_f = \left(f_R - f_T \right) = -\frac{v}{\lambda} \ \ (Hz) \tag{6}$$

where

f_R is the received frequency of the signal

f_T is the transmitted frequency of the signal

v is the relative velocity of the transmitter and receiver

λ is the wavelength of the transmitted signal

As illustrated in Fig. 11, the current velocity of a transmitter is mapped onto the LOS to a receiver that produces a measured Doppler shift. As the location and velocity of the transmitter and the Doppler shift is known, the only unknown is the location of the fixed receiver. At a later time another Doppler shift is made from a different transmitter location and these two observations can be used to resolve a 2D location (assuming: perfect frequency references on both the transmitter and receiver, stationary receiver, and no frequency distortion between transmitter and receiver). It should be noted that when dealing with a transmitter and receiver at large distance from one another (as in GNSS) the LOS is not very sensitive to the location of the receiver so precise position is hard to estimate from a single epoch of Doppler data [3]. For short-range Doppler system, where the distance between receivers and transmitters is within an order of magnitude of the desired location precision, Doppler location is far more precise.

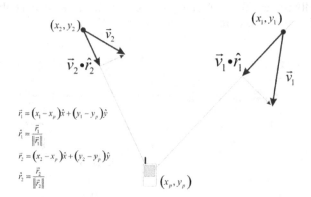

Fig. 11. Doppler velocity location

One interesting approach using Doppler location determination as presented in [24] utilizes a mobile node that transmits a signal at a known frequency and the Doppler shifted signal is measured by fixed "infrastructure" nodes. The speed of the tracked node relative to all infrastructure nodes can be calculated and used to determine both the velocity and the location of the node. The desired hardware in this approach had low computational power (8 MHz, 8-bit microprocessor) and no specialized Doppler measurement capability. They presented an approach, shown in Fig. 12, that had the tracked node transmit at a frequency f_t and one of the infrastructure nodes generates a signal f_a such that $f_t > f_a$ and when the two signals interfere with one another they produce a signal with an envelope frequency of $f_t - f_a$. This frequency difference can be measured with very fine precision using simple direct sampling. They employed a frequency difference of 300-400 Hz to enable tracking with computationally-constrained hardware. The processor sampled the radio RSSI (Received Signal Strength Indicator) which allows the measurement of the interference envelope at 8.9kHz. They used an observation window of 450 samples or approximately 18 periods of the interference envelope of 300-400 Hz which was a tradeoff between smoothing measurement noise and capturing the dynamic motion of the tracked mobile node. Results of simulation and experiment showed results of 1.3-2.2m and 0.1-0.4m/s in an operational area of 1500m^2. The limitation of this approach is that each tracked node needs to use a specific frequency (or share it in a synchronized time-sliced method) and thus cannot support an unlimited number of nodes.

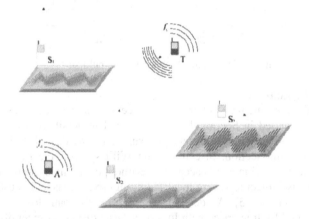

Fig. 12. Interference-based Doppler velocity and location

The advantage of the Doppler systems is that they are relatively simple in overall design. LOS requirements are the same as for any RF-based system, the frequency choice is fairly wide since it is not limited to a particular modulation or bandwidth however, processing of higher frequencies can require specialized hardware unless approaches such as the interference-based model are used.

2.4 Signal Strength

A substantial body of work has been produced examining position determination techniques based on signal strength and other signal-related properties such as signal-to-noise ratio (SNR), bit error rate (BER), Link Quality Indicator (LQI), Response Rate (RR), or carrier-to-noise ratio (C/N$_0$). The two most prevalent RF sources used in the studies are WiFi access points (IEEE 802.11) and cellular towers (both GSM and CDMA).

There are basically two general classes that signal strength solutions break down into: distance estimation (leading to lateration) and pattern recognition (also known as fingerprinting).

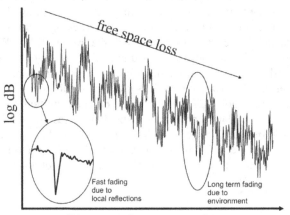

Distance from transmitter

Fig. 13. Propagation Path Loss with multipath

2.4.1 Distance Estimation

As the distance between the transmitter and the receiver increases, the RSS decreases according to as discussed in Section 1.2 and provided in detail from [5,15]. In this way RSS can be equivocated to distance measurements as described in TOA and TDOA in the following section. In cellular and WiFi systems the RSS is used for handoff and traffic processing and is accessible without changing system architecture. Operating on RSS is subject to severe local fading in urban areas and is usually unreliable in many instances [15]. As depicted in Fig. 13, the path loss of the signal strength is dominated by free-space path loss over large distance (generally hundreds of signal wavelengths) but can observe large variations in 10s of dB in shorter lengths (typically less than a few wavelengths) due to short-term fading. Long-term fading (from a few wavelengths to a hundred or so) can see fading due to reflections in the local environment such as buildings and refraction of indirect paths in the atmosphere [5]. As a specific example of a system developed using signal strength is the RADAR system developed by Microsoft Research that used 802.11 Access Points (APs) to estimate the location of a mobile transmitters within an office environment [25]. They implemented this system in two phases, first, they collected signal strength measurement data as a function of location throughout one floor of an office building to build a model (the "training phase"). They then used this model to compute location (the "test phase"). When discussing indoor location systems, it is necessary to model the transmission properties through walls and other obstructions [28]. Also relevant is the relative orientation of the receiving antenna and transmitter can greatly affect the received signal strength [25]. For near-body applications, the torso and to a lesser extent the head can impact signal strength due to absorption of certain RF frequencies especially prevalent in the microwave frequencies (1-3 GHz). The effects of the body absorption are mostly systematic and result in biases in either test or training data as compared to multipath and other scattering which have less tendency to bias. For this reason the Microsoft team collected not only location data during the training phase but orientation as well. They were able to show location errors around 2 to 3 meters for the office environment, in many cases good enough to isolate the mobile transmitter to a room however the drawback to their approach are the training phase (also requiring re-training if the physical layout should change) and also subject to environmental factors such as the number of people in the office building at a given time [25]. Recent work in ray tracing techniques have shown some promise to lower (or eliminate) the need for training in signal strength approaches providing adequate data is available on the target operational area, but the overall accuracy compared to timing-based approaches is generally lower [26]. Once the range is estimated, the range can be used in the same fashion as the timing-based methods discussed in the next section. Not only signal strength can be used for range estimation, but also Bit Error Rate (BER), Link Quality (LQ) or Response Rate (RR). For example [18] discusses a method where the RR (the percentage of times that a given transmitter was heard in all of the receiver scans at a specific distance from that transmitter) is used to estimate range for isolation of a Bluetooth device to a particular room on an office floor.

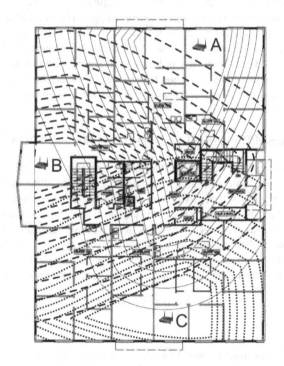

Fig. 14. Location From Signal Strength from multiple basestations

2.4.2 Pattern Recognition

In contrast to approaches that seek to map signal strength into range, pattern recognition techniques have been applied to signal metrics to form location estimates. This type of approach is often referred to as "fingerprinting" and uses previously stored measurements or calculations to map locations for later matching versus measurements. This approach takes advantage of the local-area fading characteristics by using these features from multiple transmitters to uniquely identify a location and, generally speaking, the finer the training grid the better the resultant accuracy [26,31]. Obviously the first crucial step is formation of the radio map (or training) where they can be formed by either empirical or model-based approach as shown in Fig. 14. As described in [10], in the empirical approach the data can be collected as single values at a given location or as a probability distribution. Alternatively, model-based methods can be categorized based on parameters, signal propagation and in terms of representation. In a parameter-based approach parameters can either be provided *a-priori* as in [25] or estimated from a small set of estimation fingerprints as in [28]. The propagation model can either be a direct-path only approach [25] or where multiple paths are used such as in ray tracing [26]. Finally the representation can be single values [29] or in terms of probability distributions [30].

Pattern-matching solutions have been employed on a variety of scales, on the meter-level within a building [27] to city or larger range at lower overall accuracies [31](again, mostly limited to the training set resolution). An important factor about path-loss and pattern-matching in signal strength approaches is that they are not particularly sensitive to the frequency employed. Of course the previous restrictions apply as far as attenuation and LOS issues with different frequencies but if the signal can be measured at the receiver, it can be used in signal strength approach. Another strong suit of this approach is it is not restrictive on the number receivers that can make use of the signal measurements which allows for a large number of users for a given infrastructure.

2.5 Time, Phase and Differential Timing (TOA, POA and TDOA)

GNSS systems use timing measurements collected at a receiver, which is referred to as pseudorange or the time of signal flight between the satellite transmitter and receiver corrupted by atmospheric distortion, multipath and uncertainties of the receiver clock. Precision GNSS instruments actually report the carrier phase of the signal (which have similar distortions to pseudorange). These represent Time-of-Arrival (TOA) and Phase-of-Arrival (POA) systems respectively and are not exclusively limited to satellite-based navigation systems. Typically multiple TOA ranges are combined via multi-lateration[1]. The range measurement are represented by [36]:

$$r_i(t) = \rho_i(t) + cb_i(t) \qquad (7)$$

---

[1] Multi-lateration is sometimes referred to as trilateration or tri-lateration to indicate that three range measurements at a minimum are required to determine a 2D position without ambiguity. Multi-lateration is chosen to indicate using as many ranges or pseudoranges as required. Many times this is referred to as triangulation which also describes angulation (AOA and DF) in addition to lateration [34].

In GNSS systems this is well-known as a pseudorange of $r_i(t)$ between the receiver and the i-th transmitter. The geometric distance between the transmitter and receiver is given by $\rho_i(t)$, c is the speed of light and $b_i(t)$ is the combined clock offset between the receiver and transmitter from a reference time (such as GPS time), often referred to as the *clock bias*. When examining the three-dimensional (3D) location the receiver the geometric range is:

$$\rho_i(t) = \sqrt{\left(X_i(t) - X_p\right)^2 + \left(Y_i(t) - Y_p\right)^2 + \left(Z_i(t) - Z_p\right)^2} \tag{8}$$

where $X_i(t), Y_i(t),$ and $Z_i(t)$ are the 3D location coordinates of the i-th transmitter and $X_p, Y_p,$ and Z_p is the receiver location at time t. The clock bias is a combination of the receiver clock offset from a reference time as well as the i-th transmitter's clock offset. For GNSS where the satellites are the transmitters, the satellites transmit clock parameters that allow for the correction of the satellite clock (within a small residual error) at the receiver leaving only the receiver clock offset to compute along with the position. For POA systems (Fig. 16), the same formulations as for TOA apply, however, there are usually carrier-cycle ambiguity resolution issues and if these cannot be resolved then phase measurements are used to smooth TOA measurements [25,32]. One of the best advantages of POA systems is the precision of the carrier phase. In GPS for example the codephase cycle length is 299.7 km (293 m per C/A code chip) in range where the carrier cycle length is 19 cm. This enables far more precision in

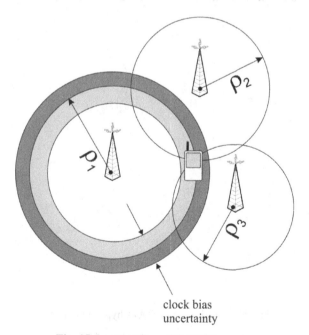

clock bias
uncertainty

Fig. 15. Location From TOA/POA ranges

ranging measurements at the expense of difficultly in resolving integer ambiguities [3,32]. For example, if there is an uncertainty of 1 m of the initial position estimate there could be as much as 10 integers for a given satellite and 10^n combinations for n satellites which could easily lead to millions of possible combinations.

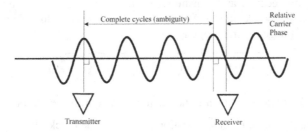

Fig. 16. Carrier Phase Ambiguity

In contrast to TOA, time-difference systems use arrival times from two transmitters at a receiver. Fig. 17 shows that the time differences in the TDOA system represent Line-of-Position hyperbolic curves along which the receiver lies. By using a second difference measurement the two-dimensional position can be determined; as shown in Fig. 17, Point B. In the case of points A and A', a single set of differences may not be able to resolve the ambiguity regarding another difference measurement [23]. A significant advantage of TDOA over TOA is that the receiver clock bias is not an important factor in location determination.

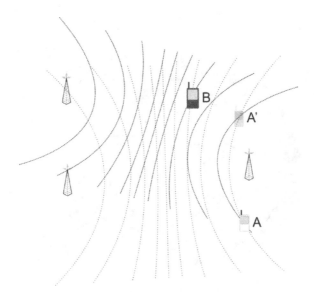

Fig. 17. Location From TDOA hyperbolic traces

3 Conclusion

The level of interest in ubiquitous location determination capability continues to increase for a variety of applications. This paper gave an overview of RF-based approaches that form the basis for all the systems employed today. These systems will continue to be key components in location solutions for years to come.

References

1. Mannings, R.: Ubiquitous Positioning. Artech House, Norwood (2008)
2. Orr, R.J., Abowd, G.D.: The Smart Floor: A Mechanism for Natural User Identification and Tracking Graphics. In: Visualisation and Usability (GVU) Centre, Georgia Institute of Technology (January 2008),
   http://www.cc.gatech.edu/fce/pubs/floor-short.pdf
3. Enge, P., Misra, P.: Global Positioning System, Signals Measurements and Performance. Ganga-Jamuna Press, Lincoln (2004)
4. Kraus, J.D., Marhefka, R.J.: Antennas: For All Applications. Tata McGraw-Hill Publishing Company Limited, New Delhi (2003)
5. Lee, W.C.Y.: Wireless and Cellular Telecommunications, 3rd edn. The McGraw-Hill Companies, New York (2006)
6. Goldhirsh, J., Vogel, W.J.: Propagation effects for land mobile satellite systems: Overview of experimental and modeling results. In: Technical Report Reference Publication 1274, NASA (1992)
7. Enge, P., et al.: Terrestrial Radionavigation Technologies. Navigation: Journal of The Institute of Navigation 42(1) (1995); Special Issue, Institute of Navigation, Fairfax VA
8. Kolodziej, K., Hjelm, J.: Local Positioning Systems: LBS Applications and Services. CRC Press Taylor & Francis Group, Boca Raton (2006)
9. International Telecommunication Union Radio Communications Sector (ITU-R),
   http://www.itu.int/net/about/itu-r.aspx
10. Kjærgaard, M.B.: A taxonomy for radio location fingerprinting. In: Hightower, J., Schiele, B., Strang, T. (eds.) LoCA 2007. LNCS, vol. 4718, pp. 139–156. Springer, Heidelberg (2007)
11. Ho, W., Smailagic, A., Siewiorek, D.P., Faloutsos, C.: An Adaptive Two-phase Approach to WiFi Location Sensing. In: IEEE Int. Conf., March 2006, vol. 5, p. 456 (2006)
12. Barcelo, F., Evennou, F., de Nardis, L., Tome, P.: Advances in Indoor Location, TOPO-CONF-2006-024 (2006)
13. Paschalidis, I., Guo, D., Lai, W., Ray, S.: Statistical Location Detection. In: Mao, G., Fidan, B. (eds.) Localization Algorithms and Strategies for Wireless Sensor Networks: Monitoring and Surveillance Techniques for Target Tracking, IGI Global (2009)
14. Ubisense (2009), http://www.ubisense.net
15. Pahlavan, K., Levesque, A.H.: Wireless Information Networks. John Wiley & Sons, Hoboken (2005)
16. Hatami, A.: Application of Channel Modeling for Indoor Localization Using TOA and RSS, PhD thesis, Department of Electrical and Computer Engineering, Worcester Polytechnic University (2006)
17. Gollu, A.: RTLS and Yard Management: Increasing Visibility, PINC Solutions presentation at the Wireless Communications Alliance RFID SIG (2007),
   http://www.wca.org/event_archives/2007/GOLLU_RFID_SEPT2007.pdf

18. Bargh, M.S., de Groote, R.: Indoor Localization Based on Response Rate of Bluetooth Inquiries. In: The First ACM International Workshop on Mobile Entity Localization and Tracking in GPS-less Environments: MELT 2008, September 19 (2008),
http://www2.parc.com/isl/projects/MELT08/program.htm
19. Aksu, A., Kabara, J., Spring, M.B.: Reduction of Location Estimation Error using Neural Networks. In: The First ACM International Workshop on Mobile Entity Localization and Tracking in GPS-less Environments: MELT (September 19, 2008),
http://www2.parc.com/isl/projects/MELT08/program.htm
20. LaMarca, A., Chawathe, Y., Consolvo, S., Hightower, J., Smith, I., Scott, J., Sohn, T., Howard, J., Hughes, J., Potter, F., Tabert, J., Powledge, P.S., Borriello, G., Schilit, B.N.: Place lab: Device positioning using radio beacons in the wild. In: Gellersen, H.-W., Want, R., Schmidt, A. (eds.) PERVASIVE 2005. LNCS, vol. 3468, pp. 116–133. Springer, Heidelberg (2005)
21. Jenkins, H.H.: Small-Aperture Radio Direction-Finding. Artech House, Norwood (1991)
22. Kayton, M., Fried, W.R.: Avionics Navigation Systems. John Wiley & Sons, Chichester (1997)
23. Enge, P., et al.: Terrestrial Radionavigation Technologies. Navigation: Journal of The Institute of Navigation 42(1) (1995); Special Issue, Institute of Navigation
24. Kusý, B., Lédeczi, À., Koutsoukos, X.: Tracking Mobile Nodes Using RF Doppler Shifts. In: ACM 5th Conference on Embedded Networked Sensor Systems (SenSys), Sydney, Australia (November 2007)
25. Bahl, P., Padmanabhan, V.N.: RADAR: An In-building RF-based User Location and Tracking System. In: IEEE INFOCOM 2000, vol. 2, pp. 775–784 (March 2000)
26. Hatami, A.: Application of Channel Modeling for Indoor Localization Using TOA and RSS, PhD thesis, Department of Electrical and Computer Engineering, Worcester Polytechnic University (May 2006)
27. King, T., Haenselmann, T., Effelsberg, W.: Deployment, Calibration, and Measurement Factors for Position Errors in 802.11-Based Indoor Positioning Systems. In: Hightower, J., Schiele, B., Strang, T. (eds.) LoCA 2007. LNCS, vol. 4718, pp. 17–34. Springer, Heidelberg (2007)
28. Ji, Y., Biaz, S., Pandey, S., Agrawal, P.: ARIADNE: a dynamic indoor signal map construction and localization system. In: MobiSys 2006: Proceedings of the 4th international conference on Mobile systems, applications and service. ACM Press, New York (2006)
29. Laitinen, H., Lahteenmaki, J., Nordstrom, T.: Database Correlation Method for GSM Location. In: Vehicular Technology Conference, VTC 2001. IEEE VTS 53rd, vol. 4, pp. 2504–2508 (2001)
30. Madigan, D., Einahrawy, E., Martin, R.P., Ju, W.H., Krishnan, P., Krishnakumar, A.S.: Bayesian indoor positioning systems. In: Proceedings IEEE INFOCOM 2005, 24th Annual Joint Conference of the IEEE Computer and Communications Societies. IEEE, Los Alamitos (2005)
31. Chen, M., Sohn, T., Chmelev, D., Haehnel, D., Hightower, J., Hughes, J., LaMarca, A., Potter, F., Smith, I., Varshavsky, A.: Practical Metropolitan-scale Positioning for GSM Phones. In: Proceedings of Ubicomp (2006)
32. Hofmann-Wellenhof, B., Lichtenegger, H., Collins, J.: Global Positioning Systems Theory and Practice, 5th revised edn. Springer, Wien (2001)
33. Federal Communications Commission (FCC), First Report and Order in The Matter of Revision of Part 15 of the Commission's Rules Regarding Ultrawideband Transmission Systems, ET-Docket 98-153, FCC 02-048 (released, April 2002)
34. Hightower, J., Borriello, G.: Location Systems for Ubiquitous Computing. IEEE Computer 34(8), 57–66 (2001)

A Survey on Localization for Mobile Wireless Sensor Networks

Isaac Amundson and Xenofon D. Koutsoukos

Institute for Software Integrated Systems (ISIS)
Department of Electrical Engineering and Computer Science
Vanderbilt University
Nashville, TN 37235, USA
isaac.amundson@vanderbilt.edu

Abstract. Over the past decade we have witnessed the evolution of wireless sensor networks, with advancements in hardware design, communication protocols, resource efficiency, and other aspects. Recently, there has been much focus on *mobile* sensor networks, and we have even seen the development of small-profile sensing devices that are able to control their own movement. Although it has been shown that mobility alleviates several issues relating to sensor network coverage and connectivity, many challenges remain. Among these, the need for position estimation is perhaps the most important. Not only is localization required to understand sensor data in a spatial context, but also for navigation, a key feature of mobile sensors. In this paper, we present a survey on localization methods for mobile wireless sensor networks. We provide taxonomies for mobile wireless sensors and localization, including common architectures, measurement techniques, and localization algorithms. We conclude with a description of real-world mobile sensor applications that require position estimation.

1 Introduction

Wireless sensor network (WSN) applications typically involve the observation of some physical phenomenon through sampling of the environment. Mobile wireless sensor networks (MWSNs) are a particular class of WSN in which mobility plays a key role in the execution of the application. In recent years, mobility has become an important area of research for the WSN community. Although WSN deployments were never envisioned to be fully static, mobility was initially regarded as having several challenges that needed to be overcome, including connectivity, coverage, and energy consumption, among others. However, recent studies have been showing mobility in a more favorable light [1]. Rather than complicating these issues, it has been demonstrated that the introduction of mobile entities can resolve some of these problems [2]. In addition, mobility enables sensor nodes to target and track moving phenomena such as chemical clouds, vehicles, and packages [3].

One of the most significant challenges for MWSNs is the need for localization. In order to understand sensor data in a spatial context, or for proper navigation

R. Fuller and X.D. Koutsoukos (Eds.): MELT 2009, LNCS 5801, pp. 235–254, 2009.
© Springer-Verlag Berlin Heidelberg 2009

throughout a sensing region, sensor position must be known. Because sensor nodes may be deployed dynamically (i.e., dropped from an aircraft), or may change position during run-time (i.e., when attached to a shipping container), there may be no way of knowing the location of each node at any given time. For static WSNs, this is not as much of a problem because once node positions have been determined, they are unlikely to change. On the other hand, mobile sensors must frequently estimate their position, which takes time and energy, and consumes other resources needed by the sensing application. Furthermore, localization schemes that provide high-accuracy positioning information in WSNs cannot be employed by mobile sensors, because they typically require centralized processing, take too long to run, or make assumptions about the environment or network topology that do not apply to dynamic networks.

This paper presents a survey and taxonomy of localization methods for mobile wireless sensor networks. Localization is a well-studied problem in several areas including robotics, mobile ad hoc and vehicular networks, and wireless sensor networks. Here, we focus solely on those methodologies that relate directly to MWSNs. In order to understand localization in the context of mobility, we begin with an overview on mobile wireless sensor networks. The overview includes common MWSN architectures, discusses the advantages of adding mobility, and describes differences with WSNs. We then provide a taxonomy of localization methods in MWSNs. In addition, we discuss the impact centralized processing and the environment have on MWSN localization. Finally, we describe real-world MWSN applications that require position estimation.

At present, the most widely used method for localization is the NAVSTAR Global Positioning System (GPS) [4]. The system consists of approximately 24 satellites that orbit the planet, of which four are required to obtain location information (3 to determine 3D position, and 1 to resolve local clock uncertainty). The satellites continuously transmit messages that contain ephemeris data, transmission time, and vital statistics. Mobile receivers are then able to compute their location using lateration based on signal time of flight and orbital position data. Commercial-use GPS is accurate to within 10 meters, is free to use anywhere on the planet and, for many mobile applications, is an ideal localization technology that should be taken advantage of. However, there are also several situations in which it will not work reliably. Because GPS requires line of sight to multiple satellites, mobile sensor networks that are deployed in urban environments, indoors, underground, or off-planet will not be able to use it. Furthermore, although GPS receivers are available for mote-scale devices, they are still relatively expensive, and therefore undesirable for many deployments. Therefore, in this survey, localization is presented from a GPS-*less* perspective (i.e., one that does not rely solely on GPS technology).

The survey is organized as follows. In Section 2, we provide a description of the key features of mobile wireless sensor networks. Section 3 then focuses specifically on localization in MWSNs, and includes a taxonomy of localization methods. Section 4 continues with a description of MWSN applications that require localization. Section 5 concludes.

2   Mobile Wireless Sensor Networks

In this section, we provide a brief taxonomy of MWSNs, including the differences between MWSNs and WSNs, and the advantages of adding mobility.

2.1   MWSN Architectures

Mobile sensor networks can be categorized by flat, 2-tier, or 3-tier hierarchical architectures [5], as illustrated in Figure 1, and described below.

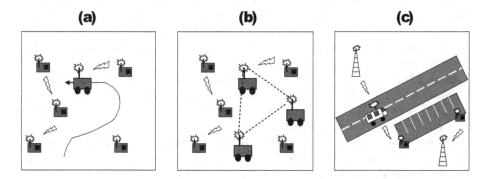

Fig. 1. (a) Flat, (b) 2-Tier, and (c) 3-Tier MWSN architectures

A flat, or planar, network architecture comprises a set of heterogeneous devices that communicate in an ad hoc manner. The devices can be mobile or stationary, but all communicate over the same network. Basic navigation systems such as [6] have a flat architecture, as pictured in Figure 1a.

The two-tier architecture consists of a set of stationary nodes, and a set of mobile nodes. The mobile nodes form an overlay network or act as data mules to help move data through the network. The overlay network can include mobile devices that have greater processing capability, longer communication range, and higher bandwidth. Furthermore, the overlay network density may be such that all nodes are always connected, or the network can become disjoint. When the latter is the case, mobile entities can position themselves in order to re-establish connectivity, ensuring network packets reach their intended destination. The NavMote system [7] takes this approach. The 2-tier architecture is pictured in Figure 1b.

In the three-tier architecture, a set of stationary sensor nodes pass data to a set of mobile devices, which then forward that data to a set of access points. This heterogeneous network is designed to cover wide areas and be compatible with several applications simultaneously. For example, consider a sensor network application that monitors a parking garage for parking space availability. The sensor network (first tier) broadcasts availability updates to compatible mobile devices (second tier), such as cell phones or PDAs, that are passing by. In turn, the cell phones forward this availability data to access points (third tier), such as cell towers, and the data are uploaded into a centralized database server. Users wishing to locate an available parking space can then access the database. The 3-tier architecture is pictured in Figure 1c.

At the node level, mobile wireless sensors can be categorized based on their role within the network:

Mobile Embedded Sensor. Mobile embedded nodes do not control their own movement; rather, their motion is directed by some external force, such as when tethered to an animal [8] or attached to a shipping container [9]. Typical embedded sensors include [10], [11], and [12].

Mobile Actuated Sensor. Sensor nodes can also have locomotion capability (for example, [13], [14], [15]), which enables them to move throughout a sensing region [6]. With this type of controlled mobility, the deployment specification can be more exact, coverage can be maximized, and specific phenomena can be targeted and followed.

Data Mule. Oftentimes, the sensors need not be mobile, but they may require a mobile device to collect their data and deliver it to a base station. These types of mobile entities are referred to as *data mules* [16]. It is generally assumed that data mules can recharge their power source automatically.

Access Point. In sparse networks, or when a node drops off the network, mobile nodes can position themselves to maintain network connectivity [16], [17]. In this case, they behave as network access points.

2.2   Advantages of Adding Mobility

Sensor network deployments are often determined by the application. Nodes can be placed in a grid, randomly, surrounding an object of interest, or in countless other arrangements. In many situations, an optimal deployment is unknown until the sensor nodes start collecting and processing data. For deployments in remote or wide areas, rearranging node positions is generally infeasible. However, when nodes are mobile, redeployment is possible. In fact, it has been shown [17], [18] that the integration of mobile entities into WSNs improves coverage, and hence, utility of the sensor network deployment. This enables more versatile sensing applications as well [1]. For example, Figure 2 illustrates a mobile sensor network that monitors wildfires. The mobile sensors are able to maintain a safe distance from the fire perimeter, as well as provide updates to fire fighters that indicate where that perimeter currently is.

In networks that are sparse or disjoint, or when stationary nodes die, mobile nodes can maneuver to connect the lost or weak communication pathways. This is not possible with static WSNs, in which the data from dead or disconnected nodes would simply be lost. Similarly, when network sinks are stationary, nodes closer to the base station will die sooner, because they must forward more data messages than those nodes further away. By using mobile base stations, this problem is eliminated, and the lifetime of the network is extended [19].

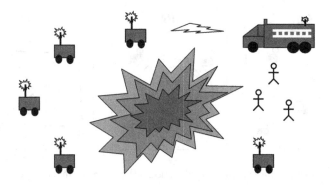

Fig. 2. A MWSN that monitors wildfires. As the fire spreads, the mobile sensors can track it, as well as stay out of its way.

Mobility also enables greater channel capacity and maintains data integrity by creating multiple communication pathways, and reducing the number of hops messages must travel before reaching their destination [20].

2.3   Differences between WSNs and MWSNs

In order to focus on the mobility aspect of wireless sensor networks, it is important to first understand how the common assumptions regarding statically-deployed WSNs change when mobile entities are introduced.

Localization. In statically deployed networks, node position can be determined once during initialization. However, those nodes that are mobile must continuously obtain their position as they traverse the sensing region. This requires additional time and energy, as well as the availability of a rapid localization service.

Dynamic Network Topology. Traditional WSN routing protocols [21], which describe how to pass messages through the network so they will most likely reach their destination, typically rely on routing tables or recent route histories. In dynamic topologies, table data become outdated quickly, and route discovery must repeatedly be performed at a substantial cost in terms of power, time, and bandwidth. Fortunately, there is an active area of research dedicated to routing in mobile ad hoc networks (MANETs), and MWSNs can borrow from this work [22].

Power Consumption. Power consumption models [23] differ greatly between WSNs and MWSNs. For both types of networks, wireless communication incurs a significant energy cost and must be used efficiently. However, mobile entities require additional power for mobility, and are often equipped with a much larger energy reserve, or have self-charging capability that enables them to plug into the power grid to recharge their batteries.

Network Sink. In centralized WSN applications, sensor data is forwarded to a base station, where it can be processed using resource-intensive methods. Data routing and aggregation can incur significant overhead. Some MWSNs use mobile base stations [19], which traverse the sensing region to collect data, or

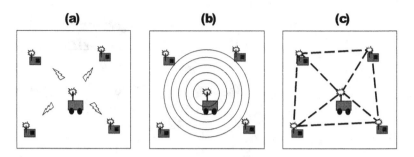

Fig. 3. Localization phases: (a) coordination, (b) measurement, and (c) position estimation

position themselves so that the number of transmission hops is minimized for the sensor nodes.

3   Localization in MWSNs

In this section, we provide a taxonomy of localization methods for MWSNs, as well as survey selected works representative of common MWSN localization. An extensive library of WSN localization research has been published within the past decade [24], [25], and many of these techniques can be applied to MWSNs. The localization techniques use diverse hardware, algorithms and signal modalities, which can be categorized along several different dimensions. We start by describing the three phases typically used in localization ([26], [27], [28]): (1) *coordination*, (2) *measurement*, and (3) *position estimation*. We then focus on other aspects of MWSN-based localization, such as the effects of mobility, centralized versus distributed processing, and the environment.

MWSN localization is typically performed as illustrated in Figure 3. A group of nodes coordinate to initiate localization. One or more nodes then emit a signal, and some property of the signal (e.g. arrival time, phase, signal strength, etc.) is observed by one or more receivers. Node position is then determined by transforming signal measurements into position estimates by means of a localization algorithm. In order to determine position, it is often necessary to enlist the help of cooperating sensor nodes that have been deployed into the environment at known positions a priori. These devices are referred to as *anchor, infrastructure,* or *seed* nodes. For example, in GPS, the infrastructure nodes are the satellites that orbit the planet. The position estimate may be relative to a set of stationary anchor nodes at known positions in a local coordinate system, or absolute coordinates may be obtained if the positions of the anchor nodes are known with respect to some global coordinate system (i.e., using GPS).

3.1   Coordination Phase

Prior to signal transmission, nodes participating in the localization typically coordinate with one another. Such coordination can include notification that the localization process is about to begin, and clock synchronization, which enables received signal data to be analyzed within a common timeframe. Coordination techniques such as reference broadcast synchronization (RBS) [29] and elapsed time on arrival (ETA) [30] exist that encapsulate both notification and synchronization into a single message. These coordination methods have microsecond accuracy and require transmission of only a single message. For example, the SyncEvent, one of the ETA primitives, declares a time in the future to begin the localization process. Encoded in the message is the timestamp of the message sender (typically the localization coordinator), which is inserted into the message immediately before transmission, thus reducing the amount of non-deterministic latency involved in the synchronization. All nodes within broadcast range will receive the message at approximately the same time instant, and assuming a negligible transit time of the radio signal through air, will be able to transform the sender timestamp into their local timescale. This technique is used in several localization schemes, including [31], [32], [9], and [6].

3.2   Measurement Phase

The measurement phase typically involves the transmission of a signal by at least one node, followed by signal processing on the other participating nodes.

Signal Modalities. The choice of signal modality used by sensor nodes is important for accurate localization, and depends on node hardware, the environment, and the application. Because WSNs are developed to provide inexpensive wide-area observation capability, it is generally undesirable to add additional hardware to the sensor board, because this increases cost and power consumption. Localization schemes will also perform differently in different environments. In humid environments, for example, radio signals perform worse than acoustic signals because moisture in the air absorbs and reflects the high frequency radio waves but does little to affect the vibrational sound waves. Finally, the application itself places some constraints on signal modality. A military application, for example, in which nodes must localize under stealth conditions, would be much better off using a silent modality such as radio frequency, rather than an audible one such as acoustic.

The acoustic modality typically employs either ultrasound or audible wave propagation. Several techniques have been published for each. Two early and commonly cited ultrasound localization techniques are Active Bats [33] and Cricket [34]. A more recent ultrasound approach, which includes a survey on ultrasonic positioning systems and challenges can be found in [35]. In the audible acoustic band, several novel localization systems have been developed, including beamforming [36], a sniper detection system [37], and generalized sound source localization [38].

Infrared (IR) signal attenuation is relatively high, requiring close proximity between transmitter and receiver. This is acceptable for most indoor localization schemes, however, outdoor localization becomes difficult, not only due to proximity issues, but also because the IR signal is difficult to read in the presence

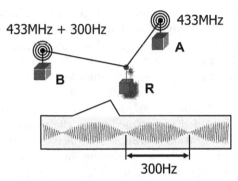

Fig. 4. Radio interferometry. Two nodes transmit a sinusoidal signal at slightly different frequencies, which interfere to create a low-frequency beat signal that can be measured using resource-constrained sensor nodes.

of sunlight. One of the earliest mobile localization systems is the Active Badge system [39], whereby a small electronic device (badge), carried by a user, emits a periodic identification signal. The signal is received by infrastructure nodes and centrally processed, allowing position information to be accessed by authorized users. Other IR localization methods can be found in [40] and [41].

Because all wireless sensor nodes have onboard radio hardware, radio frequency (RF) propagation has become a popular signal modality for localization. Signal properties such as strength, phase, or frequency are analyzed to derive range data for position estimation. One benefit of using RF is that it has been shown to achieve localization accuracy on the order of centimeters, even in sparse networks [42]. On the other hand, because typical sensor node radios transmit at frequencies between 400 MHz to 2.6 GHz, sampling the raw signal for phase or frequency cannot be done with resource-constrained hardware. Instead, methods such as radio interferometry [31] must be used to generate a low frequency beat signal, as shown in Figure 4. The frequency and phase of the beat signal can then be measured by observing the received signal strength indicator on the radio chip.

The Lighthouse [43] and Spotlight [44] localization techniques use a light beacon to determine node position. Although both methods claim high accuracy, they require line of sight, a powerful light source that will perform well in lighted areas, and customized hardware for the light source.

Measurement Techniques. Several techniques exist for obtaining bearing, range, or proximity information based on signal measurement.

The angle-of-arrival (AOA) method [45], [46], [47], [36] involves determining the angular separation between two beacons, or a single beacon and a fixed axis. By determining the AOA at a certain number of sensor nodes, position can be determined by angulation methods, as outlined in Section 3.3.

Localization by time-of-arrival (TOA) [48], [49], [50] measures the time a signal takes to arrive at some number of sensors. This requires knowing the time the signal was transmitted, and assumes tight time synchronization between sender and receiver. The signal will have known propagation properties, such as speed through air at sea level. The main drawback of this approach is that it is difficult to precisely record the arrival time of radio signals, since they travel close to the speed of light. Therefore, it works best with an acoustic source. In addition, after transmitting the signal, the source must also make its transmit time known, incurring additional communication overhead. This can be avoided by employing a round-trip TOA method [51], whereby Node A transmits a signal to Node B. Upon signal reception, Node B transmits a signal back, and Node A observes the round-trip time, accounting for deterministic delay during the communication process.

Time-difference-of-arrival (TDOA) localization [32], [38] improves upon the TOA approach by eliminating the need to know when the signal was transmitted. Several time-synchronized nodes receive a signal, and look at the difference in arrival times (or difference in signal phase) at a specific time instant. Because the signal travels at a constant speed, the source position can easily be determined if there are a sufficient number of participating nodes.

Another localization method examines the received signal strength (RSS) of a message broadcast from a known location [52], [53]. Since the free-space signal strength model is governed by the inverse-square law, accurate localization is possible. Furthermore, this typically does not involve any hardware modifications because most chips (e.g. RF, IR, etc.) provide software access to the amplitude of the received signal. Another use for RSS is *profiling* [54], [55], in which a map of RSS values is constructed during an initial training phase. Sensors then estimate their position by matching observed RSS values with the training data.

Recently, there have been several published techniques that determine the position of a node based on the observed frequency of a signal [9], [6], [56], [57]. Signal frequency will undergo Doppler-shift when the transmitter and receiver are moving relative to one another. The observed Doppler-shift at multiple infrastructure nodes can be used to derive the position and velocity of the mobile node.

The above techniques provide the most accurate position estimates, however, it is oftentimes sufficient to only localize to a region. Such a region might be a room in a house, a floor in an office building, or a city block. This type of localization can be proximity-based, such as a node is located in Region A if an anchor in Region A detects it there. Another technique to localize using hop count [58]. Because the approximate transmission range of the node radio is known, observing the number of message hops to a set of anchor nodes will constrain the target node to a specific region.

3.3 Localization Phase

The signal data obtained the measurement phase can be used to determine the approximate position of the target node. Common localization techniques for MWSNs are based on ranging, whereby distance or angle approximations are obtained. Because range data are often corrupted by noisy signal measurements, optimization methods are employed to filter the noise and arrive at a more accurate position estimate.

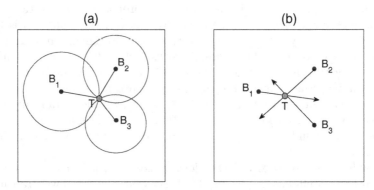

Fig. 5. The position of a target node (T) is estimated based on the known positions of beacons (B_i) using (a) lateration or (b) angulation

Lateration. When ranges between landmarks and the mobile node can be determined, lateration is used to estimate position [59]. Figure 5a illustrates the method. For two-dimensional localization, three range measurements from known positions are required. Each range can be represented as the radius of a circle, with the anchor node situated at the center. Without measurement noise, the three circles would intersect at exactly one point, the location of the target node. However, in the presence of noise, the three circles will overlap, and the target node will likely (but not necessarily) be contained within that region.

Angulation. When anchor bearings or angular separation between anchors and the mobile node can be obtained, angulation can be used to determine the position of the mobile node [60], [61], [46], [62]. This is pictured in Figure 5b. For *tri*-angulation, when two anchors are used, the target position will be identified as the third point in a triangle of two known angles (the bearings from each anchor), and the length of one side (the distance between anchor nodes). Often more than two anchor bearings are used, and target position is determined by the intersection of all bearings, as illustrated in the figure. In the presence of measurement noise, the bearings will not all intersect at the exact same point, but will instead define a region where the target node is likely to be.

Cellular Proximity. An alternative approach is the range-free method ([39], [63], [58]), whereby a node is localized to the region in which it is detected. This method generally provides a more course-grained position estimate, and depends on the density of infrastructure nodes.

Dead Reckoning. A widely used localization technique for mobile robots is dead reckoning [3], [64], [65], [7]. Robots obtain their current velocity from wheel encoders or other means, and use this information in conjunction with the amount of time that has elapsed since the last update to derive current position and heading. The major drawback of this approach is that the position estimation accrues error over time, primarily because of noisy encoder data due to uneven surfaces, wheel slippage, dust, and other factors.

Estimation Methods. When measurement data is noisy, or the system is underdefined, state estimation methods can be used. There exist a number of estimation methods, but the two main approaches are: (1) maximum likelihood estimation (MLE) [66], which estimates the values of the state based on measured data only, and no prior information about the state is used, and (2) sequential Bayesian estimation (SBE) [67], which estimates state values based on measurements, as well as prior information.

MLE methods such as [68] and [69] find the estimates for the system state by maximizing the likelihood of the measured data. In other words, MLE picks the values of the system parameters that make the observed data "more likely" than any other values for the parameters. The data likelihood is computed using a measurement model that relates the measured data to the system state.

In SBE, the system state is iteratively estimated using the recursive Bayes rule which states that the posterior is proportional to the product of the data likelihood and the predicted prior. Such methods are used in [55], [53], and [70]. Like MLE, the data likelihood is computed using a measurement model. The solution to SBE is generally intractable and cannot be determined analytically. Optimal solutions do exist in a restrictive set of cases, such as the Kalman Filter (KF) [71] and grid-based filters. More general suboptimal solutions exist, such as Extended Kalman Filter (EKF) [72] and Particle Filters (PF) that approximate the optimal Bayesian estimation. The sequential Monte Carlo (SMC) [64] method is a PF that provides a suboptimal solution by approximating the posterior density by a set of random samples (also called particles) with associated weights. As the number of particles becomes very large, the particle filter approaches an optimal solution.

3.4 The Effect of Mobility on Localization

Typically, localization of mobile sensors is performed in order to track them, or for navigational purposes. However, when sensors are mobile, we encounter additional challenges and must develop methods to address them.

One of these challenges is localization latency. If the time to perform the localization takes too long, the sensor will have significantly changed its position since the measurement took place. For example, robot navigation requires periodic position estimates in order to derive the proper control outputs for wheel angular velocity. If the robot is traveling at 1 m/s and the localization algorithm takes 5 seconds to complete from the time the ranging measurements were taken, the robot might be 5 meters off from its intended position.

Mobility may also impact the localization signal itself. For example, the frequency of the signal may undergo a Doppler shift, introducing error into the measurement. Doppler shifts occur when the transmitter of a signal is moving relative to the receiver, as illustrated in Figure 6. The resulting shift in frequency is related to the positions and relative speed of the two nodes. mTrack [32] takes this

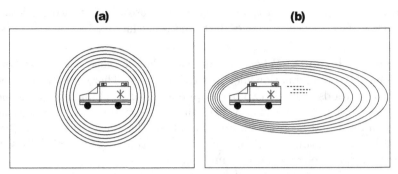

Fig. 6. (a) The frequency of a signal does not change when the transmitter and receiver are moving at the same relative speed. (b) However, when the transmitter and receiver are moving relative to one another, the signal will undergo a Doppler shift.

Doppler effect into account and uses it to refine its position estimate. Other approaches [9], [6] use the Doppler effect to directly solve for position and velocity.

If the localization technique requires line of sight (LOS), there is the possibility that the mobile sensor will move from a position with good LOS, to a position with poor LOS. When this is the case, a dense network of nodes is required to ensure there is always LOS to the mobile node, wherever it may move.

3.5   Centralized vs. Distributed Algorithms

The resource constraints inherent in WSNs pose a challenge when it comes to executing certain localization algorithms, because they require extensive memory and processor bandwidth, especially when dealing with a large number of sensors, or when using complex statistical methods to estimate range or position [25]. A centralized localization algorithm runs on a base station, and all participating nodes must forward their measurement data to the base station. The advantage of the centralized approach is an algorithm can be designed that has more accuracy, precision, and can process greater amounts of data. On the other hand, base station processing suffers from the common pitfalls of centralization, such as poor scalability, single point of failure, data routing complexity, and greater power consumption (especially for nodes closer to the base station).

When nodes are mobile, the decision to use centralized or distributed processing becomes even more important. Mobility requires continuous and rapid localization. Although centralized localization techniques exist for mobile sensors [32], [9], they are usually not fast enough for certain applications, such as navigation. For example, mTrack [32] reports a latency of approximately 5 seconds. dNav [6], on the other hand, is distributed, and takes less than 1 second on average to return position and velocity estimates.

3.6   The Impact of Environment on Localization

The environment plays a significant role in the effectiveness of a localization method. As a result, there is no one localization method that will be accurate for all situations. Different environmental factors are listed below, as well as the effect they have on the aforementioned localization methods.

Ambient temperature, pressure, and humidity can affect localization accuracy, because these directly impact the crystal oscillator in the transceiver. Furthermore, it has been well established that radio wave propagation is affected by precipitation, including moisture in the air, therefore localization techniques that use RF measurements can be impaired under these conditions [73]. One of the biggest problems with GPS is that it does not work reliably under water, indoors, or even when it is cloudy. This is because the GPS receiver requires line of sight to up to four satellites orbiting the planet [4].

At present there is a major effort underway to develop accurate localization methods in indoor environments. Indoor applications that require node position estimation are challenging because most propagation methods and measurement techniques suffer from multipath effects [74], where obstacles (e.g. walls, furniture, people, etc.) cause signal reflections that interfere with each other. In addition, many of the existing localization techniques that provide good accuracy outdoors, will not work indoors.

4 MWSN Applications with Localization Requirements

Although MWSNs are still in their infancy, several types of applications have already been developed in which localization plays an integral part. The applications fall under four main categories, (a) commercial, (b) environmental, (c) civil, and (d) military, however, most span more than just one of these.

4.1 Commercial

As MWSNs grow in popularity, we expect to see a burst of applications in the commercial sector that require some kind of position data.

- **Service Industry.** One such area is the service industry. Companies such as Skilligent [75] are developing software protocols for service robots that perform tasks such as basic patient care in nursing homes, maintenance and security in office buildings, and food and concierge service in restaurants and hotels. All of these applications require a mechanism for position estimation. Skilligent uses a visual localization system based on pattern matching. Objects are used as landmarks, and are loaded into the system a priori, or dynamically at runtime. The robot learns its position by matching video images with landmark information.
- **Housekeeping.** The iRobot Roomba [76] is an automated vacuum cleaning robot for domestic use. The Roomba creates a map of the room as it moves by using feedback from a variety of bumper and optical sensors. Wheel encoders provide run-time position information that enable it to cover the entire room. The Roomba also uses a self-docking station to automatically recharge its batteries.

4.2 Environmental

MWSNs have become a valuable asset for environmental monitoring. This is thanks in part to their ability to be deployed in remote areas and for their ability to gather data of wide areas of interest.

– **Wildlife Tracking.** ZebraNet [8] is an early MWSN, in which mote-scale wireless devices were fitted to zebras for the purpose of tracking their movement. Due to the remote region, there was no cellphone coverage, so data was routed through the peer-to-peer network to mobile base stations. The zebras were not constrained to certain areas, and other than the small devices attached to their bodies, left undisturbed. To accomplish this level of tracking without the use of MWSNs would not be possible.
– **Pollution Monitoring.** A mobile air quality monitoring system is presented in [77]. Sensor nodes that measure specific pollutants in the air are mounted on vehicles. As the vehicles move along the roadways, the sensors sample the air, and record the concentration of various pollutants along with location and time. When the sensors are in the proximity of access points, the data are uploaded to a server and published on the web.

4.3 Civil

One of the areas that has great potential for MWSN utility is that of civil services. This includes those non-military municipal applications that keep society running efficiently and safely.

– **Pothole Detection.** In [78], a system is developed to detect potholes on city streets. Deployed on taxi cabs, the sensor nodes contain an accelerometer, and can communicate using either opportunistic WiFi or cellular networks.
– **Wireless E-911.** In North America, the Enhanced 911 emergency telecommunications service, or *E911* [79], was established to connect callers with emergency services in a manner that would associate a physical location with the phone number of the caller. *Wireless* E-911 is the second phase of the E911 service mandated by the FCC, which requires wireless cellular devices to automatically provide user location when the service is invoked. This is an important requirement, however, its implementation is non-trivial, and different carriers choose to use different methods, including embedded GPS chips, and multilateration and angulation based on the known locations of cell towers.

4.4 Military / Aerospace

One of the biggest promoters, as well as one of the biggest funders, of wireless sensor technology is the military. There is a clear interest in localization services, tracking friendly and hostile entities, and navigation of autonomous robots, and intensive research is carried out in this area.

– **Shooter Detection / Weapon Classification.** In [80], a soldier-wearable sensor system is developed that not only identifies the location of an enemy sniper, but also identifies the weapon being fired. Each sensor consists of an array of microphones mounted on the helmet of a soldier. The sensor observes both the shock wave of the projectile, as well as the muzzle blast from the weapon, and based on TDOA, as well as properties of the acoustic signal, is able to triangulate the enemy position and classify the weapon type.

- **Autonomous Deployment.** In [81] an unattended aerial vehicle is used for sensor network deployment and repair. Such deployments aid the military in battlefield surveillance and command and control field operations.

5   Conclusion

In this paper, we presented a survey and taxonomy on localization for mobile wireless sensor networks. Localization in MWSNs entails new challenges that result from integrating resource-constrained wireless sensors on a mobile platform. The localization methods and algorithms that provide greater accuracy on larger-footprint mobile entities with fewer resource restrictions are no longer applicable. Similarly, centralized and high-latency localization techniques for static WSNs are undesirable for the majority of MWSN applications.

There are several directions for future work in MWSN localization. Reducing localization latency is one of the most important benchmarks for MWSNs. Currently, a tradeoff exists between the rapid execution of an algorithm and its accuracy. Additional work is needed that focused on reducing run-time latency, while maintaining positioning accuracy. In addition, the majority of localization algorithms to date are centralized. For mobile sensor localization, this is often a poor design choice, due to the additional latency and energy costs incurred. The development of more distributed localization techniques would be a welcome addition to MWSN localization. There is much interest in localization in urban and indoor areas where obstacles such as vehicles, walls, people, and furniture cause multipath propagation and loss of line of sight. Most current methods use some variation of RSS profiling, in conjunction with optimization techniques. However, new methods are required as we expand mobile sensing to areas where training data cannot safely be obtained, such as urban war zones or burning buildings. Lastly, mobile actuated sensors are now being developed with mote-sized form factors. Like embedded sensor nodes, these devices also have resource constraints, which limit their ability to navigate a sensing region in the same way a robot with a full array of sensors and powerful processing capability might. We can expect to see many advances in mobile sensor navigation in the near future.

Acknowledgements. This work was supported in part by ARO MURI grant W911NF-06-1-0076, NSF grant CNS-0721604, and NSF CAREER award CNS-0347440. The authors also wish to thank Manish Kushwaha for his valuable input.

References

1. Ekici, E., Gu, Y., Bozdag, D.: Mobility-based communication in wireless sensor networks. Communications Magazine, IEEE 44(7), 56–62 (2006)
2. Munir, S.A., Ren, B., Jiao, W., Wang, B., Xie, D., Ma, J.: Mobile wireless sensor network: Architecture and enabling technologies for ubiquitous computing. In: Proceedings of the 21st International Conference on Advanced Information Networking and Applications Workshops, AINAW (2007)
3. Tilak, S., Kolar, V., Abu-Ghazaleh, N.B., Kang, K.D.: Dynamic localization control for mobile sensor networks. In: Proceedings of the IEEE International Workshop on Strategies for Energy Efficiency in Ad Hoc and Sensor Networks (2005)
4. Hofmann-Wellenhof, B., Lichtenegger, H., Collins, J.: Global Positioning System: Theory and Practice, 4th edn. Springer, Heidelberg (1997)
5. Munir, S.A., Ren, B., Jiao, W., Wang, B., Xie, D., Ma, J.: Mobile wireless sensor network: Architecture and enabling technologies for ubiquitous computing. In: International Conference on Advanced Information Networking and Applications Workshops, vol. 2, pp. 113–120 (2007)
6. Amundson, I., Koutsoukos, X., Sallai, J.: Mobile sensor localization and navigation using RF doppler shifts. In: 1st ACM International Workshop on Mobile Entity Localization and Tracking in GPS-less Environments, MELT (2008)
7. Fang, L., Antsaklis, P.J., Montestruque, L., Mcmickell, M.B., Lemmon, M., Sun, Y., Fang, H., Koutroulis, I., Haenggi, M., Xie, M., Xie, X.: Design of a wireless assisted pedestrian dead reckoning system – the NavMote experience. In: IEEE Transactions on Instrumentation and Measurement, vol. 54(6), pp. 2342–2358 (2005)
8. Juang, P., Oki, H., Wang, Y., Martonosi, M., Peh, L., Rubenstein, D.: Energy-efficient computing for wildlife tracking: Design tradeoffs and early experiences with zebranet. In: Proc. of ASPLOS-X (2002)
9. Kusý, B., Lédeczi, A., Koutsoukos, X.: Tracking mobile nodes using RF doppler shifts. In: SenSys 2007: Proceedings of the 5th international conference on Embedded networked sensor systems, pp. 29–42. ACM, New York (2007)
10. Crossbow MICAz (MPR2400) Radio Module,
    http://www.xbow.com/Products/productsdetails.aspx?sid=101
11. Dutta, P., Grimmer, M., Arora, A., Bibyk, S., Culler, D.: Design of a wireless sensor network platform for detecting rare, random, and ephemeral events. In: Proc. of IPSN/SPOTS (April 2005)
12. Polastre, J., Szewczyk, R., Culler, D.: Telos: Enabling ultra-low power wireless research. In: Proc. of IPSN/SPOTS (April 2005)
13. Dantu, K., Rahimi, M., Shah, H., Babel, S., Dhariwal, A., Sukhatme, G.S.: Robomote: enabling mobility in sensor networks. In: The Fourth International Symposium on Information Processing in Sensor Networks, IPSN (2005)
14. Friedman, J., Lee, D.C., Tsigkogiannis, I., Wong, S., Chao, D., Levin, D., Kaisera, W.J., Srivastava, M.B.: Ragobot: A new platform for wireless mobile sensor networks. In: International Conference on Distributed Computing in Sensor Systems, DCOSS (2005)
15. Bergbreiter, S., Pister, K.S.J.: CotsBots: An off-the-shelf platform for distributed robotics. In: Proceedings of the IEEE/RSJ International Conference on Intelligent Robots and Systems, IROS (2003)
16. Shah, R., Roy, S., Jain, S., Brunette, W.: Data mules: modeling a three-tier architecture for sparse sensor networks. In: Proceedings of the First IEEE International Workshop on Sensor Network Protocols and Applications (2003)

17. Wang, G., Cao, G., Porta, T., Zhang, W.: Sensor relocation in mobile sensor networks. In: IEEE INFOCOM (2005)
18. Liu, B., Brass, P., Dousse, O., Nain, P., Towsley, D.: Mobility improves coverage of sensor networks. In: Proceedings of the 6th ACM international symposium on Mobile ad hoc networking and computing (MobiHoc), pp. 300–308 (2005)
19. Gandham, S., Dawande, M., Prakash, R., Venkatesan, S.: Energy efficient schemes for wireless sensor networks with multiple mobile base stations. In: IEEE Global Telecommunications Conference, GLOBECOM (2003)
20. Kansal, A., Somasundara, A.A., Jea, D.D., Srivastava, M.B., Estrin, D.: Intelligent fluid infrastructure for embedded networks. In: Proceedings of the 2nd international conference on Mobile systems, applications, and services (MobiSys), pp. 111–124 (2004)
21. Al-Karaki, J.N., Kamal, A.E.: Routing techniques in wireless sensor networks: a survey. IEEE Wireless Communications 11(6), 6–28 (2004)
22. Abolhasan, M., Wysocki, T., Dutkiewicz, E.: A review of routing protocols for mobile ad hoc networks. Ad Hoc Networks 2(1), 1–22 (2004)
23. Wang, Q., Hempstead, M., Yang, W.: A realistic power consumption model for wireless sensor network devices. In: 3rd Annual IEEE Communications Society on Sensor and Ad Hoc Communications and Networks (SECON), vol. 1, pp. 286–295 (2006)
24. Hightower, J., Borriello, G.: Location systems for ubiquitous computing. IEEE Computer 34(8), 57–66 (2001)
25. Mao, G., Fidan, B., Anderson, B.D.O.: Wireless sensor network localization techniques. Computer Networks 51(10), 2529–2553 (2007)
26. Brooks, R.R., Griffin, C., Friedlander, D.S.: Self-organized distributed sensor network entity tracking. The International Journal of High Performance Computing Applications 16(3) (2002)
27. Moore, D., Leonard, J., Rus, D., Teller, S.: Robust distributed network localization with noisy range measurements. In: SenSys 2004: Proceedings of the 2nd international conference on Embedded networked sensor systems, pp. 50–61 (2004)
28. Girod, L., Lukac, M., Trifa, V., Estrin, D.: The design and implementation of a self-calibrating acoustic sensing platform. In: Proc. of ACM SenSys (November 2006)
29. Elson, J., Girod, L., Estrin, D.: Fine-grained network time synchronization using reference broadcasts. SIGOPS Oper. Syst. Rev. 36(SI), 147–163 (2002)
30. Kusý, B., Dutta, P., Levis, P., Maróti, M., Lédeczi, A., Culler, D.: Elapsed time on arrival: a simple and versatile primitive for canonical time synchronization services. International Journal of Ad Hoc and Ubiquitous Computing 2(1) (2006)
31. Maróti, M., Kusý, B., Balogh, G., Völgyesi, P., Nádas, A., Molnár, K., Dóra, S., Lédeczi, A.: Radio interferometric geolocation. In: Proc. of ACM SenSys (November 2005)
32. Kusý, B., Sallai, J., Balogh, G., Lédeczi, A., Protopopescu, V., Tolliver, J., DeNap, F., Parang, M.: Radio interferometric tracking of mobile wireless nodes. In: Proc. of MobiSys (2007)
33. Harter, A., Hopper, A., Stegglesand, P., Ward, A., Webster, P.: The anatomy of a context-aware application. In: Mobile Computing and Networking, pp. 59–68 (1999)
34. Priyantha, N.B., Chakraborty, A., Balakrishnan, H.: The Cricket location-support system. In: Proc. of MobiCom (August 2000)
35. McCarthy, M., Duff, P., Muller, H.L., Randell, C.: Accessible ultrasonic positioning. IEEE Pervasive Computing 5(4), 86–93 (2006)

36. Chen, J., Yao, K., Hudson, R.: Source localization and beamforming. Signal Processing Magazine, IEEE 19(2), 30–39 (2002)
37. Lédeczi, A., Nádas, A., Völgyesi, P., Balogh, G., Kusý, B., Sallai, J., Pap, G., Dóra, S., Molnár, K., Maróti, M., Simon, G.: Countersniper system for urban warfare. ACM Transactions on Sensor Networks 1(1), 153–177 (2005)
38. Williams, S.M., Frampton, K.D., Amundson, I., Schmidt, P.L.: Decentralized acoustic source localization in a distributed sensor network. Applied Acoustics 67 (2006)
39. Want, R., Hopper, A., Falcao, V., Gibbons, J.: The active badge location system. ACM Transactions on Information Systems 40 (1992)
40. Brassart, E., Pegard, C., Mouaddib, M.: Localization using infrared beacons. Robotica 18(2), 153–161 (2000)
41. Kemper, J., Linde, H.: Challenges of passive infrared indoor localization. In: Proceedings of 5th Workshop on Positioning, Navigation and Communication (WPNC), pp. 63–70 (2008)
42. Kusý, B., Balogh, G., Völgyesi, P., Sallai, J., Nádas, A., Lédeczi, A., Maróti, M., Meertens, L.: Node-density independent localization. In: Proc. of IPSN/SPOTS (April 2006)
43. Römer, K.: The lighthouse location system for smart dust. In: Proceedings of the 1st International Conference on Mobile Systems, Applications and Services (MobiSys), pp. 15–30 (2003)
44. Stoleru, R., He, T., Stankovic, J.A., Luebke, D.: A high-accuracy, low-cost localization system for wireless sensor networks. In: Proceedings of ACM SenSys (November 2005)
45. Bekris, K.E., Argyros, A.A., Kavraki, L.E.: Angle-based methods for mobile robot navigation: Reaching the entire plane. In: International Conference on Robotics and Automation (2004)
46. Niculescu, D., Nath, B.: Ad hoc positioning system (APS) using AOA. In: Proceedings of the Twenty-Second Annual Joint Conference of the IEEE Computer and Communications Societies, INFOCOM (2003)
47. Friedman, J., Charbiwala, Z., Schmid, T., Cho, Y., Srivastava, M.: Angle-of-arrival assisted radio interferometry (ARI) target localization (2008)
48. Priyantha, N.B., Balakrishnan, H., Demaine, E.D., Teller, S.: Mobile-assisted localization in wireless sensor networks. In: Proceedings of the IEEE 24th Annual Joint Conference of the IEEE Computer and Communications Societies (INFOCOM), vol. 1 (2005)
49. Caffery Jr., J.J.: A new approach to the geometry of toa location. In: 52nd Vehicular Technology Conference, 2000. IEEE VTS-Fall VTC 2000, vol. 4, pp. 1943–1949 (2000)
50. Wang, X., Wang, Z., O'Dea, B.: A toa-based location algorithm reducing the errors due to non-line-of-sight (nlos) propagation. IEEE Transactions on Vehicular Technology 52(1), 112–116 (2003)
51. Günther, A., Hoene, C.: Measuring round trip times to determine the distance between WLAN nodes. In: Boutaba, R., Almeroth, K.C., Puigjaner, R., Shen, S., Black, J.P. (eds.) NETWORKING 2005. LNCS, vol. 3462, pp. 768–779. Springer, Heidelberg (2005)
52. Lee, H., Wicke, M., Kusy, B., Guibas, L.: Localization of mobile users using trajectory matching. In: MELT 2008: Proceedings of the first ACM international workshop on Mobile entity localization and tracking in GPS-less environments, pp. 123–128 (2008)

53. Madigan, D., Einahrawy, E., Martin, R., Ju, W.H., Krishnan, P., Krishnakumar, A.: Bayesian indoor positioning systems. In: INFOCOM 2005. Proceedings of IEEE 24th Annual Joint Conference of the IEEE Computer and Communications Societies, March 2005, vol. 2, pp. 1217–1227 (2005)

54. Bahl, P., Padmanabhan, V.N.: Radar: An in-building RF-based user-location and tracking system. In: Proc. IEEE INFOCOM, March 2000, vol. 2, pp. 775–784 (2000)

55. Ladd, A., Bekris, K., Rudys, A., Wallach, D., Kavraki, L.: On the feasibility of using wireless ethernet for indoor localization. IEEE Transactions on Robotics and Automation 20(3), 555–559 (2004)

56. Ledeczi, A., Volgyesi, P., Sallai, J., Thibodeaux, R.: A novel RF ranging method. In: Sixth Workshop on Intelligent Solutions in Embedded Systems, WISES (2008)

57. Chang, H.I., Tian, J.B., Lai, T.T., Chu, H.H., Huang, P.: Spinning beacons for precise indoor localization. In: Proceedings of the Sixth ACM conference on Embedded network sensor systems, SenSys (2008)

58. Niculescu, D., Nath, B.: Dv based positioning in ad hoc networks. Journal of Telecommunication Systems 22, 267–280 (2003)

59. Manolakis, D.: Efficient solution and performance analysis of 3-D position estimation by trilateration. IEEE Transactions on Aerospace and Electronic Systems 32(4), 1239–1248 (1996)

60. Esteves, J., Carvalho, A., Couto, C.: Generalized geometric triangulation algorithm for mobile robot absolute self-localization (2003)

61. McGillem, C., Rappaport, T.: A beacon navigation method for autonomous vehicles. IEEE Transactions on Vehicular Technology 38(3), 132–139 (1989)

62. Betke, M., Gurvits, L.: Mobile robot localization using landmarks 13(2), 251–263 (April 1997)

63. Shang, Y., Ruml, W., Zhang, Y., Fromherz, M.: Localization from connectivity in sensor networks. IEEE Transactions on Parallel and Distributed Systems 15(11), 961–974 (2004)

64. Hu, L., Evans, D.: Localization for mobile sensor networks. In: Proceedings of the 10th annual international conference on Mobile computing and networking, MobiCom (2004)

65. Zhang, P., Martonosi, M.: Locale: Collaborative localization estimation for sparse mobile sensor networks. In: Proceedings of the 7th international conference on Information processing in sensor networks, IPSN (2008)

66. Kay, S.M.: Fundamentals of Statistical Signal Processing, Volume I: Estimation Theory. Prentice-Hall, Englewood Cliffs (1993)

67. Arulampalam, M., Maskell, S., Gordon, N., Clapp, T.: A tutorial on particle filters for online nonlinear/non-gaussian bayesian tracking. IEEE Transactions on Signal Processing 50(2), 174–188 (2002)

68. Kuang, X., Shao, H.: Maximum likelihood localization algorithm using wireless sensor networks. In: First International Conference on Innovative Computing, Information and Control (ICICIC), vol. 3, pp. 263–266 (2006)

69. Mendalka, M., Kulas, L., Nyka, K.: Localization in wireless sensor networks based on zigbee platform. In: 17th International Conference on Microwaves, Radar and Wireless Communications (MIKON), pp. 1–4 (2008)

70. Fox, V., Hightower, J., Liao, L., Schulz, D., Borriello, G.: Bayesian filtering for location estimation. Pervasive Computing, IEEE 2(3), 24–33 (2003)

71. Kalman, R.E.: A new approach to linear filtering and prediction problems. Trans. ASME, Journal of Basic Engineering (1960)

72. Welch, G., Bishop, G.: An introduction to the Kalman filter. Technical Report TR 95-041, Department of Computer Science, University of North Carolina at Chapel Hill (2004)
73. Sizun, H.: Radio wave propagation for telecommunication applications. Springer, Heidelberg (2004)
74. Hashemi, H.: The indoor radio propagation channel. Proceedings of the IEEE 81(7), 943–968 (1993)
75. Skilligent: Skilligent visual localization system, http://www.skilligent.com/products/robot-navigation.shtml
76. iRobot: Roomba vacuum cleaning robot, http://www.irobot.com
77. Völgyesi, P., Nádas, A., Koutsoukos, X., Lédeczi, A.: Air quality monitoring with sensormap. In: Proceedings of the 7th international conference on Information processing in sensor networks (IPSN), pp. 529–530 (2008)
78. Eriksson, J., Girod, L., Hull, B., Newton, R., Madden, S., Balakrishnan, H.: The pothole patrol: using a mobile sensor network for road surface monitoring. In: Proceedings of the 6th international conference on Mobile systems, applications, and services, MobiSys (2008)
79. Federal Communications Commission: Enhanced 911, http://www.fcc.gov/pshs/services/911-services/enhanced911/
80. Völgyesi, P., Balogh, G., Nádas, A., Nash, C., Lédeczi, A.: Shooter localization and weapon classification with soldier-wearable networked sensors. 5th International Conference on Mobile Systems, Applications, and Services, MobiSys (2007)
81. Corke, P., Hrabar, S., Peterson, R., Rus, D., Saripalli, S., Sukhatme, G.: Autonomous deployment and repair of a sensor network using an unmanned aerial vehicle. In: IEEE International Conference on Robotics and Automation, pp. 3602–3609 (2004)

Performance of TOA- and RSS-Based Indoor Geolocation for Cooperative Robotic Applications

Nader Bargshady, Nayef A. Alsindi, and Kaveh Pahlavan

Center for Wireless Information Network Studies
Worcester Polytechnic Institute
Worcester, MA. 01609, USA
{nbargsha,Kaveh}@wpi.edu
http://www.cwins.wpi.edu

Abstract. Recently, *cooperative* robotic applications have attracted considerable attention. Cooperative assignments for robots demand accurate localization. Since in an indoor environment localization using GPS does not render a satisfactory result, we need to resort to different approaches for indoor geolocation. Precise localization information means better coordination that enables us to manipulate robots more effectively for variety of tasks. In this paper, the cooperative localization performance accuracy for a multi-robot operation is examined using empirical models for ranging estimates in an indoor environment scenario at the third floor of the Atwater Kent Laboratory (AKL) in the Worcester Polytechnic Institute. The two widely used ranging techniques are *Time Of Arrival* (TOA) using Ultra-wideband (UWB) and *Received Signal Strength* (RSS) using WiFi signals. We use empirical statistical models for UWB TOA-based and WiFi RSS-based operations in order to determine the *Cramér-Rao-Lower-Bound* (CRLB) on the performance of localization techniques in our multi-robot operation scenarios. We determine the performance of the localization of robots when they are localized individually versus when they are benefited from *cooperative localization.*

1 Introduction

The emergence of cooperative robotic applications calls for variety of tasks that require more precise coordination in order to effectively manipulate the robots for either a single or joint task operation. In robotic applications, vision modality is used for finding the coordinate information[1],[2]. The vision modality requires a Line-Of-Sight (LOS) condition and in most indoor environments there are walls, partitions and furniture that block the view and create a Non-Line-Of-Sight (NLOS) situation. A loss of visual data results in severe degradations in localization precision. Using the radio propagation signals and models we can potentially overcome this problem and achieve better localization in the absence of visual data. The traditional RF localization is performed by GPS signals,

R. Fuller and X.D. Koutsoukos (Eds.): MELT 2009, LNCS 5801, pp. 255–266, 2009.
© Springer-Verlag Berlin Heidelberg 2009

which does not work properly in indoor environment. As a result, recently non-GPS localization using other opportunistic signals have attracted considerable attention[3],[4]. The most popular RF indoor geolocation systems use Time Of Arrival (TOA) of the UWB signals and Received Signal Strength (RSS) of WiFi signals[5]. The TOA-based UWB signals provide for more precise localization but the coverage is limited and the design needs new hardware infrastructure. The RSS-based WiFi localization can be implemented in software and on the existing wide spread WiFi infrastructure with a significantly wider coverage than UWB systems. As a result, UWB localization has found its way in wireless sensor networks[6] and the WiFi localization is used for both indoor and outdoor applications[5]. The results of quantitatively comparative performance evaluation of UWB and WiFi localization in [7] reveals that RSS WiFi localization provides a statistically smooth but less reliable localization while the UWB's localization in most occasions provides more precise localization than the RSS localization. To resolve these difficulties in precise RF localization techniques, cooperative localization has offered itself as a solution for applications in wireless sensor networks[5],[6],[7].

In this paper, we analyze the merits of RF cooperative localization using TOA-based UWB and RSS-based WiFi technologies for cooperative robotic applications in our *modeling and simulation* environment. In our simulation environment we take advantage of empirical results obtained from third floor of AKL for portion of our modeling (TOA) to be described in Section 2. We define a movement scenario for multi-robot operation based on the layout of the third floor of AKL (Fig. 2) with respect to four *static* reference points. With the use of our mix-mode (empirical and theoretical) modeling we derive *Cramér-Rao-Lower-Bound* (CRLB) for calculation of the localization error for individual robot and when they operate in a cooperative manner both for UWB and WiFi systems. We assume the robots to be equipped with UWB or WiFi in our simulation environment.

Localization error consists of ranging error and positioning error. For UWB systems we use empirical ranging error models for TOA-based systems reported in [8],[9]. For WiFi ranging error we use the IEEE 802.11 channel model for calculation of the RSS and the CRLB for RSS links presented in [10]. For performance evaluation, we examine the relative performance of the two approaches by applying the CRLB for cooperative localization presented in [8],[11].

Section 2 describes our models for ranging error in each link. Section 3 presents the CRLB for localization used in this paper. Section 4 describes the multi-robot operation scenario. In Section 5 we provide the comparative performance evaluation results and in Section 6 we conclude this paper.

2   Models for Link Errors for TOA and RSS

In this section we describe the two TOA- and RSS-based models that provide us with variance of the link's ranging error as a function of distance between two RF radiating sources.

Remark 1. A source is defined as a reference point (UWB or WiFi) or a robot (equipped with UWB or WiFi). In our simulation, we use four static reference points to localize (trilateration) the robots. We further simulate the result of cooperation among robots by calculating the distance between the robots to achieve more accuracy in localization.

The variance of TOA- or RSS-based model is used to calculate *Cramér-Rao-Lower-Bound* (CRLB) for our localization performance bound in Section 3.

For calculation of variance of the ranging error for RSS-based WiFi localization we use the result of derivation of CRLB for the ranging error in RSS systems from [10]:

$$\sigma_R^2 = var(\widehat{d}) \geq \left(\frac{\ln 10}{10}\right)^2 \cdot \frac{\eta^2}{\alpha^2} \cdot d^2 \qquad (1)$$

Where d is the distance between two sources, η^2 is the variance of the log-normal $\mathcal{N}(0, \eta^2)$ shadow fading of the environment, and α is the so-called distance-power gradient of the environment. Using IEEE 802.11 path-loss model[12] for our simulations, $\eta = 8$ and α takes on a value of 2 for LOS situations and 3.5 for NLOS conditions. The distance d between two sources $Si(x_i, y_i)$ and $Sj(x_j, y_j)$ is:

$$d = \sqrt{(x_j - x_i)^2 + (y_j - y_i)^2} \qquad (2)$$

For TOA based systems in the absence of multi-path the CRLB is given by[13]:

$$\sigma_D^2 \geq \frac{1}{8\pi^2} \cdot \frac{1}{SNR} \cdot \frac{1}{T \cdot W} \cdot \frac{1}{f_0^2} \cdot \frac{1}{1 + \frac{W^2}{12f_0^2}} \qquad (3)$$

Where T is the observation time, SNR is the Signal-To-Noise-Ratio, f_0 is the center frequency of operation and W is the bandwidth of the system. This bound is valid for GPS applications in the open areas and provides very small errors regardless of the distance. However, in multi-Path rich indoor environments, where direct paths between the sources are blocked, this bound is loose and researchers resort to empirical modeling of the ranging error[12]. Hence, we also resort to empirical models presented in [8],[9]. In our simulations we have used the specific model for ranging error in UWB systems presented in [8]. In this empirical model the ranging error is assumed to be a Guassian random variable whose mean and variance are functions of two power thresholds:

$$\sigma_T^2 = \begin{cases} \mathcal{N}(\mu_1, \sigma_1^2) & RSS(d) \leq Th_1 \\ \mathcal{N}(\mu_2, \sigma_2^2) & Th_1 < RSS(d) \leq Th_2 \\ \mathcal{N}(\mu_3, \sigma_3^2) & Th_2 < RSS(d) \end{cases} \qquad (4)$$

Where RSS(d) is the received power at a robot in a distance d from a reference point, $Th_{1,2}$ are the power thresholds, and μ_i and σ_i^2 are mean and variance of the ranging error which are also a function of existence of the direct paths. The thresholds used in the model are $Th_1 = -80$dBm and $Th_2 = -100$dBm and the corresponding σ_i^2 are:

$$\sigma_i^2 = \begin{cases} (0.12)^2 & RSS(d) \leq -80 \\ (0.3)^2 & -80 < RSS(d) \leq -100 \\ (1.4)^2 & -100 < RSS(d) \end{cases} \qquad (5)$$

The mean and variance for calculation of the error for different channel conditions are given in [8]. The model used for calculation of the RSS is given by:

$$RSS(d) = RSS(1) - 10 \cdot \alpha \cdot \log d - \chi \qquad (6)$$

in which $RSS(1)$ is the received signal strength at 1 meter distance from a reference point, d is a distance, and χ is the lognormal shadow fading.

In our simulation[8], $RSS(1) = -42$ (dBm) and (χ, α) take on the set values of $(\chi = 6.8dB, \alpha = 2.0)$ when in Line-Of-Sight (LOS) and $(\chi = 8.5dB, \alpha = 5.6)$ when in Non-Line-Of-Site (NLOS).

3   CRLB for Cooperative Localization

In this section we do not discuss higher level protocols or implementation issues. We merely derive the performance bound based on Cramér-Rao-Lower-Bound (CRLB). The CRLB provides a lower bound on the variance achievable by any unbiased location estimator. The bound is useful as a guideline: knowing the best an estimator (TOA- or RSS-based) can possibly do that can help us judge our approach in this section.

The derived values for *distance d and variance* $(\sigma_R^2$ *or* $\sigma_T^2)$ in *Section 2* are used in this section to calculate CRLB that allow us to assess the performance of our estimate. We describe our derivation for CRLB from papers [8],[11].

Remark 1. For further simplification we assume the the LOS and NLOS variances can coexist as part of the same diagonal matrix Λ_λ:

$$\Lambda_\lambda = \begin{pmatrix} \lambda_1 & \dots & 0 \\ \vdots & \ddots & \vdots \\ 0 & \dots & \lambda_M \end{pmatrix} \qquad (7)$$

Where M refers to number of "reference points" (UWB or WiFi), for minimum of 3 where in our case we use 4 reference points. The element $\lambda_{1:M}$ is the inverse of σ_R^2 or σ_T^2 for the corresponding $d_{1:M}$ for every robot location. The corresponding distance d from Eq. 2 is used to assemble our geometry vector with respect to our reference points:

$$\Delta_{vec} = \begin{pmatrix} \Delta_{x_1} & \dots & \Delta_{x_M} \\ \Delta_{y_1} & \dots & \Delta_{y_M} \end{pmatrix} \qquad (8)$$

Where Δ_x and Δ_y are partial derivatives of calculated distance, d with reference to x and y coordinates respectively:

$$\Delta_x = \frac{(x_i - x_j)}{d_{ij}} \qquad \Delta_y = \frac{(y_i - y_j)}{d_{ij}} \qquad . \tag{9}$$

The Fisher Information Matrix, FIM is calculated as:

$$FIM = \Delta_{vec} \cdot \Lambda_\lambda \cdot \Delta_{vec}^T \qquad . \tag{10}$$

Remark 1. FIM matrix is always *full rank* and its inverse always exists in the cases we investigated.

The Cramér-Rao-Lower-Bound(CRLB) for each *individual* link d_{ij} is derived by the inverse of FIM matrix:

$$CRLB = \left[\Delta_{vec} \cdot \Lambda_\lambda \cdot \Delta_{vec}^T \right]^{-1} \qquad . \tag{11}$$

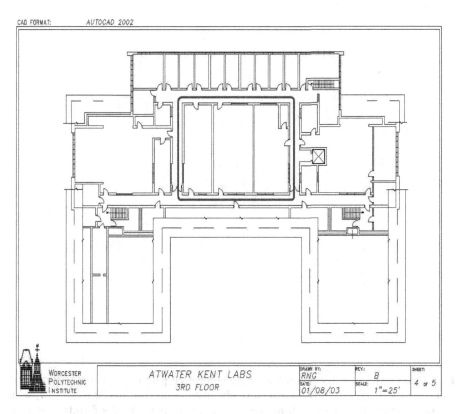

Fig. 1. ATWATER KENT LABS 3rd Floor, Worcester Polytechnic Institute the corridor chosen for the movement of robots is identified by dark solid blue rectangle. More details are shown in (Fig. 2).

Finally, evaluating the Root-Mean-Square-Error, RMSE for each *individual* link, d_{ij}:

$$RMSE^1 = \sqrt{trace\left(\left[\Delta_{vec} \cdot \Lambda_\lambda \cdot \Delta_{vec}^T\right]^{-1}\right)} \quad . \tag{12}$$

In Section 4, we analyze the CRLB results for our smulation scenario.

4   Performance Evaluation Scenario

For our performance evaluation, we define a scenario in the third floor of the AKL in Worcester Polytechnic Institute, shown in Fig. 1. In this scenario we assume the three robots are moving in the connected corridors in the central part of the third floor. This route is shown by solid blue line in the center of the building layout. Fig. 2 shows a closeup route for the robots, location of the

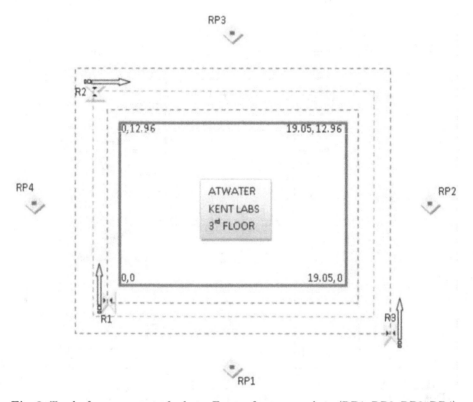

Fig. 2. Tracks for movement of robots: Four **reference points** (RP1, RP2, RP3, RP4) and *three robots* (R1, R2, R3). Each move about its respective (*dotted rectangles*) with 0.4 meter separation among them. The *three arrows* point in the direction of each robot's movement.

---

[1] "trace" stands for the trace of matrix.

reference points and the track for each robot. There are four *reference points (RP1, RP2, RP3, RP4)* and *three robots (R1, R2, R3)*. The dotted lines in Fig. 2 shows the route taken by individual robot that are 0.4 meters apart to avoid collision. The arrows in Fig. 2 show the direction and the starting point of each robot's movement. The first two robots move clockwise and the third, counterclockwise. We assume all robots start at the same time and move at the same speed. The reference points are located in the center of each side of the Route 2.6 meters away from the central track of the robot number one (R1). In our performance evaluation scenarios we assume the reference points to be either an UWB transmitter or a WiFi access point.

For each sample of time we use the location of each robot to determine its distance from other reference points and robots when in cooperating mode. The distances are used in the equations provided in Section 2 to determine the variance of the localization error, σ_R^2 or σ_T^2, associated with UWB or WiFi links, respectfully. The variances of ranging errors obtained for appropriate links are then used for calculation of the CRLB for positioning, described in Section 3.

5   Results and Discussion

In this section we discuss the results of our simulation runs presented in *figures 3,4,5 and 6*. We compare the performance of cooperative and non-cooperative

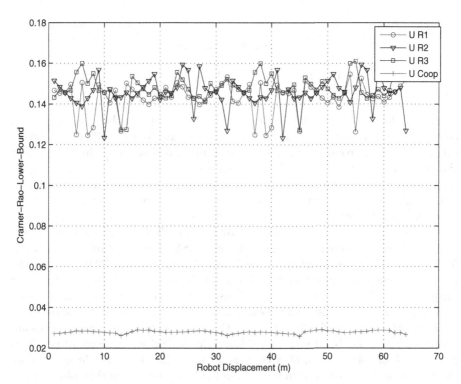

Fig. 3. UWB : *RMSE versus robot displacement*

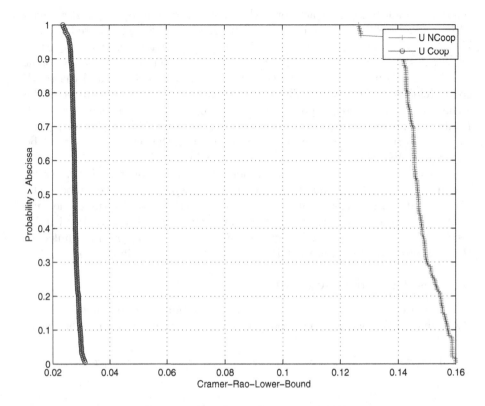

Fig. 4. UWB : *Probability versus RMSE*

operations for UWB and WiFi localization. Fig. 3 shows the performance of cooperative versus non-cooperative operation in sixty four equally distanced locations across the route when sources are using UWB signals. The lower curve shows the RMSE for variance of positioning error of all robots when they cooperate for localization. The three top plots show the RMSE of localization for each individual robot when they obtain their location from reference points only (no cooperation). As expected, in the vicinity of reference points we notice better performance, the few undershoots as shown in Fig. 3. On average, cooperation among the robots show improvement in the RMSE of localization by a factor of 5. Fig. 4 shows the commulative distribution function of the RMSE across the route. As shown on the left of Fig. 4, the RMSE is nicely confined in a narrow range $\simeq 0.03$ whereas in the individual cases we notice RMSE as high as 0.16.

In Fig. 5, shows the performance of cooperative versus non-cooperative operation in sixty four equally distanced locations across the route when sources are using WiFi signals. The lower curve shows the RMSE for variance of positioning error of all robots when they cooperate for localization. The three top plots show the RMSE of localization for each individual robot when they obtain their location from reference points in lieu of cooperation. On average cooperation

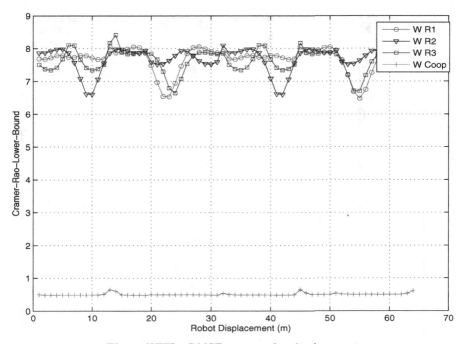

Fig. 5. WIFI : *RMSE versus robot displacement*

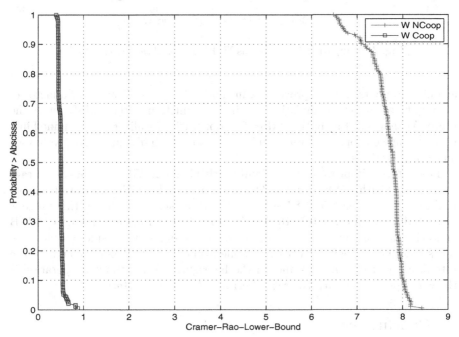

Fig. 6. WIFI : *Probability versus RMSE*

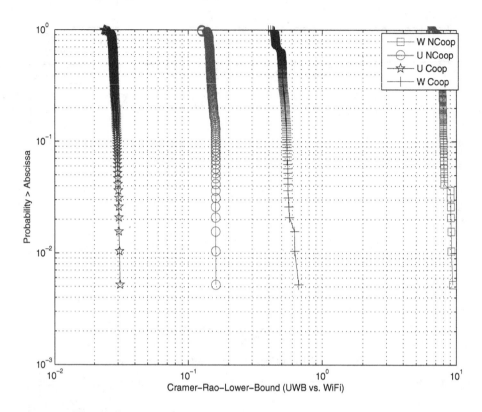

Fig. 7. Comparing UWB versus WIFI :*Probability versus RMSE*

among the robots show improvement in the RMSE of localization by a factor of 15. Fig. 6 shows the commulative distribution function of the RMSE across the route. As shown in Fig. 6, the RMSE for cooperative localization is confined in a narrow range of approximately $\simeq 0.5$ whereas in the individual cases, we notice an error as high as 8.5.

The factors of improvement are significantly higher in WiFi, 15 times on average as compared with UWB, 5 times. The range of error in WiFi localization, shown in Fig. 5, is between 6.5 to 8.5 while in UWB, as shown in Fig. 3, this range is restricted between 0.125 to 0.16. Fig. 7, shows the overall performance of UWB and WiFi in cooperative and non-Cooperative for our scenario side-by-side.

As shown in far left, UWB with cooperation is the best performer with the rate of error of 0.03 meters and on the far right, WiFi without cooperation is the worst performer with the rate of error of 6.5 to 8.5 meters.

6   Conclusions

With recent proliferation of wireless devices in robotic applications, support for localization services using radio signals has attracted tremendous attention in research community. In this paper we simulated our models and analyzed the quantitative performance of two widely used localization techniques based on TOA and RSS using UWB and WiFi transmission medium, respectively. Our results of modeling and simulation for our scenario at the third floor of the AKL showed that UWB-based localization provides error in the range of 0.125 to 0.16 meters for non-cooperative and 0.025 to 0.029 meters For cooperative localization. The WiFi localization range for non-cooperative localization was 6.5 to 8.5 meters and for cooperative 0.47 and 0.64 meters. These quantitative results provides an insight to the common believe that UWB is more accurate than WiFi both in cooperative and non-Cooperative mode. In both cases of UWB and WiFi gained significant improvement by cooperative localization using robots. However, WiFi localization benefits much higher rate of improvement through cooperation (15 times) as compared with UWB localization (5 times).

Acknowledgments. The authors would like to thank their colleagues at the Center for Wireless Information Network Studies for the helpful discussions. In particular Dr. Ning Yang for helpful discussions, carefull review and precise comments.

References

1. Spletzer, J., Das, A.K., Fierro, R., Taylor, C.J., Kumar, V., Ostrowski, J.P.: Cooperative localization and control for multi-robot manipulation. In: Proceedings of 2001 IEEE/RSJ International Conference on Intelligent Robots and Systems, 2001, 29 October-3 November 2001, vol. 2, pp. 631–636 (2001)
2. Rekleitis, I.M., Dudek, G., Milios, E.E.: Multi-robot cooperative localization: a study of trade-offs between efficiency and accuracy. In: IEEE/RSJ International Conference on Intelligent Robots and System, 2002, 30 September-5 October 2002, vol. 3, pp. 2690–2695 (2002)
3. Pahlavan, K., Akgul, F.O., Heidari, M., Hatami, A., Elwell, J.M., Tingley, R.D.: Indoor geolocation in the absence of direct path. IEEE Wireless Communications 13(6), 50–58 (2006)
4. Pahlavan, K., Li, X., Makela, J.P.: Indoor geolocation science and technology. IEEE Communications Magazine 40(2), 112–118 (2002)
5. Pahlavan, K.: WiFi and UWB RF Localization-Principles and Applications. In: Pahlavan, K. (ed.) IEEE Short Course. PIMRC 2006, Helsinki, Finland (September 2006)
6. Silverstrim, J., Passmore, R., Pahlavan, K., Sadler, B.: Wireless sensor networks with geolocation. Defense Tech. Briefs 3(3), 14–15 (2009)
7. Hatami, A., Pahlavan, K.: Performance Comparison of RSS and TOA Indoor Geolocation Based on UWB Measurement of Channel Characteristics. In: 17th Annual IEEE International Symposium on Personal Indoor and Mobile Radio Communications (PIMRC 2006), Helsinki, Finland (September 11-14, 2006)
8. Alsindi, N.A.: Cooperative Localization Bounds for Indoor Ultra-Wideband Wireless Sensor Networks. Hindawi Publishing Corporation EURASIP Journal on Advances in Signal Processing 2008, Article ID 852509, 13 pages
9. Alavi, B., Pahlavan, K.: Modeling of the TOA-based distance measurement error using UWB indoor radio measurements. Communications Letters, IEEE 10(4), 275–277 (2006)

10. Qi, Y., Kobayashi, H.: On relation among time delay and signal strength based geolocation methods. In: Global Telecommunications Conference, 2003. GLOBE-COM 2003, December 2003, vol. 7(1-5), pp. 4079–4083 (2003)
11. Savvides, A., Garber, W.L., Moses, R.L., Srivastava, M.B.: An analysis of error inducing parameters in multihop sensor node localization. IEEE Transactions on Mobile Computing 4(6), 567–577 (2005)
12. Kaveh, P., Allen, H.L.: Wireless Information Networks, 2nd edn., p. 738. Wiley & Sons, Inc., Chichester (2005)
13. Gezici, S., Tian, Z., Giannakis, G.B., Kobayashi, H., Molisch, A.F., Poor, H.V., Sahinoglu, Z.: Localization via ultra-wideband radios. In: IEEE Signal Processing Magazine (Special Issue on Signal Processing for Positioning and Navigation with Applications to Communications), July 2005, vol. 22(4), pp. 70–84 (2005)

Author Index